Housing Grants, Construction and Regeneration Act 1996

CHAPTER 53

Explanatory Notes are available separately

Housing Grants, Construction and Regeneration Act 1996

CHAPTER 53

HOUSING GRANTS, CONSTRUCTION AND REGENERATION ACT 1996

PART I

GRANTS, &C. FOR RENEWAL OF PRIVATE SECTOR HOUSING

CHAPTER I

THE MAIN GRANTS

Introductory

Preliminary conditions

Renovation grants

Grant conditions and repayment

Supplementary provisions

CHAPTER II

GROUP REPAIR SCHEMES

Introductory

Participation in group repair scheme

Variation of group repair scheme

Conditions of participation

Supplementary provisions

Registration and discipline

Miscellaneous

PART IV

GRANTS &C. FOR REGENERATION, DEVELOPMENT AND RELOCATION

Financial assistance for regeneration and development

Relocation grants in clearance areas

PART V

MISCELLANEOUS AND GENERAL PROVISIONS

Miscellaneous provisions

General provisions

SCHEDULES

SCHEDULE 1 — Private sector renewal: consequential amendments

Housing Grants, Construction and Regeneration Act 1996

1996 CHAPTER 53

An Act to make provision for grants and other assistance for housing purposes and about action in relation to unfit housing; to amend the law relating to construction contracts and architects; to provide grants and other assistance for regeneration and development and in connection with clearance areas; to amend the provisions relating to home energy efficiency schemes; to make provision in connection with the dissolution of urban development corporations, housing action trusts and the Commission for the New Towns; and for connected purposes. [24th July 1996]

Be it enacted by the Queen's most Excellent Majesty, by and with the advice and consent of the Lords Spiritual and Temporal, and Commons, in this present Parliament assembled, and by the authority of the same, as follows:—

PART I

GRANTS, &C. FOR RENEWAL OF PRIVATE SECTOR HOUSING

CHAPTER I

THE MAIN GRANTS

Introductory

1   Grants for improvements and repairs, &c

(1) Grants are available from local housing authorities in accordance with this Chapter towards the cost of works required for—

    (a)   the improvement or repair of dwellings, houses in multiple occupation or the common parts of buildings containing one or more flats,

    (b)   the provision of dwellings or houses in multiple occupation by the conversion of a house or other building, and

    (c)   the provision of facilities for disabled persons in dwellings and in the common parts of buildings containing one or more flats.

(2) A grant relating to—

    (a)   the improvement or repair of a dwelling, or

    (b)   the provision of dwellings by the conversion of a house or other building,

is referred to as a "renovation grant".

(3) A grant relating to the improvement or repair of the common parts of a building is referred to as a "common parts grant".

(4) A grant for the provision of facilities for a disabled person—

    (a)   in a dwelling, or

    (b)   in the common parts of a building containing one or more flats,

is referred to as a "disabled facilities grant".

(5) A grant for—

    (a)   the improvement or repair of a house in multiple occupation, or

    (b)   the provision of a house in multiple occupation by the conversion of a house or other building,

is referred to as an "HMO grant".

(6) In the following provisions of this Chapter the expression "grant", without more, means any of these types of grant.

2   Applications for grants

(1) No grant shall be paid unless an application for it is made to the local housing authority in accordance with the provisions of this Chapter and is approved by them.

(2) An application for a grant shall be in writing and shall specify the premises to which it relates and contain—

    (a)   particulars of the works in respect of which the grant is sought (in this Chapter referred to as the "relevant works");

    (b)   unless the local housing authority otherwise direct in any particular case, at least two estimates from different contractors of the cost of carrying out the relevant works;

    (c)   particulars of any preliminary or ancillary services and charges in respect of the cost of which the grant is also sought; and

    (d)   such other particulars as may be prescribed.

(3) In this Chapter "preliminary or ancillary services and charges", in relation to an application for a grant, means services and charges which—

    (a)   relate to the application and the preparation for and the carrying out of works, and

    (b)   are specified for the purposes of this subsection by order of the Secretary of State.

(4) The Secretary of State may by regulations prescribe a form of application for a grant and an application for a grant to which any such regulations apply is not validly made unless it is in the prescribed form.

Preliminary conditions

3   Ineligible applicants

(1) No grant is payable under this Chapter unless the applicant is aged 18 or over on the date of the application.

In the case of a joint application, any applicant under the age of 18 years on the date of the application shall be left out of account.

(2) No grant is payable under this Chapter if the person who would otherwise qualify as the applicant for the grant is—

  (a)   a local authority;

  (b)   a new town corporation;

  (c)   an urban development corporation;

  (d)   a housing action trust;

  (e)   the Development Board for Rural Wales;

  (f)   a health authority, special health authority or NHS trust;

  (g)   a police authority established under section 3 of the Police Act 1964;

  (h)   a joint authority established by Part IV of the Local Government Act 1985;

  (i)   a residuary body established by Part VII of that Act; or

  (j)   an authority established under section 10(1) of that Act (waste disposal).

(3) No grant is payable under this Chapter if the applicant is of a description excluded from entitlement to grant aid by regulations made by the Secretary of State.

(4) Regulations under subsection (3) may proceed wholly or in part by reference to the provisions relating to entitlement to housing benefit, or any other form of assistance, as they have effect from time to time.

4   The age of the property

(1) A local housing authority shall not entertain an application for a grant in respect of premises provided (by construction or conversion) less than ten years before the date of the application, unless—

  (a)   the application is for a disabled facilities grant, or

  (b)   the application is for an HMO grant in respect of a house in multiple occupation provided by conversion.

(2) The Secretary of State may by order amend subsection (1) so as to substitute another period for that specified.

5   Excluded descriptions of works

(1) No grant is payable in respect of works of a description excluded from grant aid under this Chapter by regulations made by the Secretary of State.

(2) Regulations may be made with respect to local housing authorities generally or to a particular local housing authority and may be made with respect to particular areas.

(3) Regulations may specify descriptions of works for which grant aid is not to be available without the Secretary of State's consent, which may be given—

    (a)   to local housing authorities generally or to a particular local housing authority,

    (b)   with respect to particular areas, or

    (c)   with respect to applications generally or to a particular description of application.

6   Defective dwellings

(1) No grant is payable if—

    (a)   the dwelling, house or building is or forms part of a building of a class designated under section 528 or 559 of the Housing Act 1985 (defective dwellings),

    (b)   the applicant is eligible for assistance under Part XVI of that Act in respect of a defective dwelling which is or forms part of the dwelling, house or building concerned, and

    (c)   the relevant works are, within the meaning of that Part, works required to reinstate that defective dwelling.

(2) If the local housing authority consider that the relevant works include works for which assistance is available under Part XVI of the Housing Act 1985 (assistance for owners of defective housing), they shall treat the application as if the relevant works did not include those works.

Renovation grants

7   Renovation grants: owner's applications and tenant's applications

(1) A local housing authority shall not entertain an application for a renovation grant unless they are satisfied—

    (a)   that the applicant has, or proposes to acquire, an owner's interest in every parcel of land on which the relevant works are to be carried out, or

    (b)   in the case of an application other than a conversion application, that the applicant is a qualifying tenant of the dwelling (alone or jointly with others) but does not have, or propose to acquire, an owner's interest in the dwelling.

(2) References in this Chapter to an "owner's application" or a "tenant's application", in relation to a renovation grant, shall be construed accordingly.

(3) In accordance with directions given by the Secretary of State, a local housing authority may treat the condition in subsection (1)(a) as met by a person who has, or proposes to acquire, an owner's interest in only part of the land concerned.

(4) References in this Chapter to"a qualifying owner's interest", in relation to an application for a renovation grant, are to an owner's interest meeting the condition in subsection (1)(a) or treated by virtue of subsection (3) as meeting that condition.

(5) In this Chapter a"qualifying tenant", in relation to an application for a renovation grant, means a person who (alone or jointly with others) is a tenant of the premises to which the application relates—

    (a) who is required by the terms of his tenancy to carry out the relevant works and whose tenancy is not of a description excluded from this subsection by order of the Secretary of State, or

    (b) whose tenancy is of a description specified for the purposes of this subsection by order of the Secretary of State.

(6) In subsection (5) "tenant" includes a person having a licence to occupy the premises concerned which satisfies such conditions as may be specified by order of the Secretary of State.

References in this Chapter to tenants and other expressions relating to tenancies, in the context of a tenant's application for a renovation grant, shall be construed accordingly.

8   Renovation grants: certificates required in case of owner's application

(1) A local housing authority shall not entertain an owner's application for a renovation grant unless it is accompanied by an owner-occupation certificate or a certificate of intended letting in respect of the dwelling to which the application relates or, in the case of a conversion application, in respect of each of the dwellings to be provided.

(2) An "owner-occupation certificate" certifies that the applicant—

    (a) has or proposes to acquire a qualifying owner's interest, and

    (b) intends that throughout the grant condition period he or a member of his family will live in the dwelling as his (or that member's) only or main residence.

(3) A "certificate of intended letting" certifies that the applicant—

    (a) has or proposes to acquire a qualifying owner's interest, and

    (b) intends that throughout the grant condition period the dwelling will be let or available for letting as a residence (and not for a holiday) to a person who is not connected with the owner for the time being of the dwelling.

In paragraph (b) "letting" does not include a letting on a long tenancy.

(4) In subsection (3) references to letting include the grant of a licence to occupy premises.

References in this Chapter to tenants and other expressions relating to tenancies, in the context of a certificate of intended letting, shall be construed accordingly.

(5) Where section 10 applies (prior qualifying period in certain cases) a local housing authority shall not entertain an owner's application for a renovation grant unless it is also accompanied by a certificate specifying how the requirements of that section are met.

9   Renovation grants: certificates required in case of tenant's application

(1) A local housing authority shall not entertain a tenant's application for a renovation grant unless it is accompanied by a tenant's certificate.

(2) A "tenant's certificate" certifies—

    (a) that the applicant is a qualifying tenant of the dwelling, and

(b) that he or a member of his family intends to live in the dwelling as his (or that member's) only or main residence.

(3) Except where the authority consider it unreasonable in the circumstances to require such a certificate, they shall not entertain a tenant's application for a renovation grant unless it is also accompanied by a certificate of intended letting (see section 8(3)) by the person who at the time of the application is the landlord under the tenancy.

(4) Where section 10 applies (prior qualifying period in certain cases) a local housing authority shall not entertain a tenant's application for a renovation grant unless it is also accompanied by a certificate specifying how the requirements of that section are met.

10   Renovation grants: prior qualifying period

(1) Subject to subsection (3), a local housing authority shall not entertain an application to which this section applies unless they are satisfied—
   (a) that the ownership or tenancy condition (see section 11) was met throughout the qualifying period, and
   (b) in the case of an application accompanied by an owner-occupation certificate or a tenant's certificate, that the applicant lived in the dwelling as his only or main residence throughout that period.

In the case of a joint application it is sufficient if those conditions are met by any of the applicants.

(2) The qualifying period for the purposes of this section is the period of three years, or such other period as may be specified by order of the Secretary of State, ending with the date of the application.

(3) A local housing authority may dispense with compliance with either or both of the conditions in subsection (1), and may do so either generally or in relation to particular cases or descriptions of case.

(4) Subject to subsection (5), this section applies to every application for a renovation grant, other than—
   (a) a conversion application,
   (b) an application in respect of a dwelling in a renewal area,
   (c) an application in respect of works to provide means of escape in case of fire or other fire precautions, or
   (d) an application of any other description excepted from this section by order of the Secretary of State.

(5) This section does not apply to a landlord's application unless the Secretary of State by order so provides, which he may do with respect to all landlord's applications or any description of landlord's application.

Any such order may provide that this section applies to a landlord's application notwithstanding that it is of a kind mentioned in paragraphs (a) to (d) of subsection (4).

(6) A "landlord's application" for a renovation grant means an owner's application which is accompanied by a certificate of intended letting.

A conversion application for the provision of two or more dwellings shall not be treated as a landlord's application if any of the certificates accompanying the application is an owner-occupation certificate.

11   Prior qualifying period: the ownership or tenancy condition

(1) The "ownership or tenancy condition" for the purposes of section 10 is that the applicant had a qualifying owner's interest in, or was a qualifying tenant of, the dwelling.

That condition shall be treated as having been met in the following circumstances.

(2) Where the applicant took his owner's interest or became a qualifying tenant under the will or on the intestacy of a member of his family, the ownership or tenancy condition shall be treated as having been met—

    (a)   during any period when the deceased both held a qualifying owner's interest in or was a qualifying tenant of the dwelling and lived in the dwelling as his only or main residence, and

    (b)   if immediately before his death the deceased both—

        (i)   held such an interest or was such a tenant, and

        (ii)   lived in the dwelling as his only or main residence,

    during any period not exceeding one year when his personal representatives, or the Public Trustee under section 9 of the Administration of Estates Act 1925, held such an interest or was such a tenant.

(3) The local housing authority may treat a person as continuing to meet the residence requirement in subsection (2)(a) or (b)(ii) for up to a year after he has, by reason of age or infirmity—

    (a)   gone to live with and be cared for by a member of his family, or

    (b)   gone to live in a hospital, hospice, sheltered housing, residential care home or similar institution.

(4) Where the applicant took his owner's interest or became a qualifying tenant by virtue of a disposal made by a member of his family, and the authority are satisfied—

    (a)   that the person making the disposal was elderly or infirm, and

    (b)   that he made the disposal with the intention of—

        (i)   going to live with and be cared for by a member of his family, or

        (ii)   going to live in a hospital, hospice, sheltered housing, residential care home or similar institution as his only or main residence,

    the ownership or tenancy condition shall be treated as having been met during any period ending on the date of the disposal when the person making the disposal held a qualifying owner's interest in or was a qualifying tenant of the dwelling.

(5) Where the applicant took his owner's interest or became a qualifying tenant by virtue of a disposal made by his spouse, and the authority are satisfied that the disposal was made as a result of arrangements in relation to divorce, judicial separation or declaration of nullity of marriage, the ownership or tenancy condition shall be treated as having been met during any period ending on the date of the disposal when the spouse held a qualifying owner's interest in or was a qualifying tenant of the dwelling.

(6) The references in subsection (5) to the spouse of the applicant—

    (a)   in the case of divorce, include his former spouse, and

    (b)   in the case of a declaration of nullity, shall be construed as references to the other party to the marriage.

12 Renovation grants: purposes for which grant may be given

(1) The purposes for which an application for a renovation grant, other than a conversion application, may be approved are the following—

    (a) to comply with a notice under section 189 of the Housing Act 1985 (repair notice in respect of unfit premises) or otherwise to render the dwelling fit for human habitation;

    (b) to comply with a notice under section 190 of that Act (repair notice in respect of premises not unfit but in need of substantial repair) or otherwise to put the dwelling in reasonable repair;

    (c) to provide adequate thermal insulation;

    (d) to provide adequate facilities for space heating;

    (e) to provide satisfactory internal arrangements;

    (f) to provide means of escape in case of fire or other fire precautions, not being precautions required under or by virtue of any enactment (whenever passed);

    (g) to ensure that the dwelling complies with such requirements with respect to construction or physical condition as may be specified by the Secretary of State;

    (h) to ensure that there is compliance with such requirements with respect to the provision or condition of services and amenities to or within the dwelling as are so specified;

    (i) any other purpose for the time being specified for the purposes of this section by order of the Secretary of State.

(2) The purpose for which a conversion application may be approved is to provide one or more dwellings by the conversion of a house or other building.

(3) If in the opinion of the authority the relevant works are more or less extensive than is necessary to achieve any of the purposes set out in subsection (1) or (2), they may, with the consent of the applicant, treat the application as varied so that the relevant works are limited to or, as the case may be, include such works as seem to the authority to be necessary for that purpose.

(4) The reference in paragraph (f) of subsection (1) to precautions required under or by virtue of an enactment does not include precautions required to comply with a notice under section 352 of the Housing Act 1985 (notice requiring execution of works to render house in multiple occupation fit for number of occupants) so far as it relates to premises which are not part of a house in multiple occupation for the purposes of this Part.

(5) In exercise of the powers conferred by paragraphs (g) and (h) of subsection (1) the Secretary of State may specify requirements generally or for particular cases, and may specify different requirements for different areas.

13 Renovation grants: approval of application

(1) The local housing authority may approve an application for a renovation grant if they think fit, subject to the following provisions.

(2) The authority shall not approve an application for a renovation grant unless they are satisfied that the works are necessary for one or more of the purposes set out in section 12(1) or (2).

(3) Where an authority entertain an owner's application for a renovation grant made by a person who proposes to acquire a qualifying owner's interest, they shall not approve the application until they are satisfied that he has done so.

(4) An authority proposing to approve an application for a renovation grant shall consider whether the premises to which it relates are fit for human habitation.

(5) If it appears to the authority that the premises are not fit for human habitation, they shall not approve the application unless they are satisfied—

    (a) that on completion of the relevant works, together with any other works proposed to be carried out, the premises will be fit for human habitation,

    (b) that there are satisfactory financial and other arrangements for carrying out those works, and

    (c) that the carrying out of the works is the most satisfactory course of action.

(6) In considering whether to approve an application for a renovation grant the authority shall have regard to the expected life of the building (taking account, where appropriate, of the effect of carrying out the works).

Common parts grants

14   Common parts grants: occupation of flats by occupying tenants

(1) A local housing authority shall not entertain an application for a common parts grant unless they are satisfied that at the date of the application at least the required proportion of the flats in the building concerned is occupied by occupying tenants.

(2) In this Chapter an "occupying tenant", in relation to a flat in a building, means a person who has in relation to the flat (alone or jointly with others)—

    (a) a tenancy to which section 1 of the Landlord and Tenant Act 1954 or Schedule 10 to the Local Government and Housing Act 1989 applies (long tenancies at low rents),

    (b) an assured tenancy, a protected tenancy, a secure tenancy or a statutory tenancy,

    (c) a protected occupancy under the Rent (Agriculture) Act 1976 or an assured agricultural occupancy within the meaning of Part I of the Housing Act 1988, or

    (d) a tenancy or licence which satisfies such conditions as may be specified by order of the Secretary of State,

and who occupies the flat as his only or main residence.

References in this Chapter to other expressions relating to tenancies, in the context of an application for a common parts grant, shall be construed accordingly.

(3) The "required proportion" mentioned in subsection (1) is three-quarters or such other proportion as may be—

    (a) specified for the purposes of this section by an order of the Secretary of State, or

    (b) approved by him, in relation to a particular case or description of case, on application made by the local housing authority concerned.

15   Common parts grants: landlord's and tenants' applications

(1) A local housing authority shall not entertain an application for a common parts grant unless they are satisfied—

    (a)   that the applicant has an owner's interest in the building and has a duty or power to carry out the relevant works, or

    (b)   that the application is made by at least three-quarters of the occupying tenants of the building who under their tenancies have a duty to carry out, or to make a contribution in respect of the carrying out of, some or all of the relevant works.

(2) References in this Chapter to a "landlord's application" and a "tenants' application", in relation to a common parts grant, shall be construed accordingly.

(3) In deciding whether the requirement in subsection (1)(b) is met—

    (a)   where a tenancy is held by two or more persons jointly, those persons shall be regarded as a single occupying tenant; and

    (b)   a tenant whose tenancy is of a description specified for the purposes of that paragraph by order of the Secretary of State shall be treated as an occupying tenant falling within that paragraph.

(4) A person who has an owner's interest in the building and who has a duty or power to carry out any of the relevant works may also join in a tenants' application for a common parts grant; and where such a person does join in an application, he is in this Chapter referred to as a "participating landlord".

16   Common parts grants: certificate required to accompany application

(1) A local housing authority shall not entertain a landlord's application for a common parts grant unless it is accompanied by a certificate signed by the applicant which—

    (a)   specifies the interest of the applicant in the building, and

    (b)   certifies that the required proportion of the flats in the building is occupied by occupying tenants.

(2) A local housing authority shall not entertain a tenants' application for a common parts grant unless it is accompanied by a certificate signed by each of the applicants which—

    (a)   specifies the interest of each of the applicants in each flat in the building, and

    (b)   certifies that the required proportion of the flats in the building is occupied by occupying tenants.

17   Common parts grants: purposes for which grant may be given

(1) The purposes for which an application for a common parts grant may be approved are—

    (a)   to comply with a notice under section 189 of the Housing Act 1985 (repair notice in respect of unfit premises) or otherwise to cause the building to meet the requirements in section 604(2) of that Act;

    (b)   to comply with a notice under section 190 of that Act (repair notice in respect of premises not unfit but in need of substantial repair) or otherwise to put the building in reasonable repair;

    (c)   to comply with a notice under section 352 of that Act (notice requiring works to render premises fit for the number of occupants) or otherwise to enable the house to meet one or more of the requirements in subsection (1A) of that section;

(d) to provide adequate thermal insulation;

(e) to provide adequate facilities for space heating;

(f) to provide satisfactory internal arrangements;

(g) to provide means of escape in case of fire or other fire precautions, not being precautions required under or by virtue of any enactment (whenever passed);

(h) to ensure that the building complies with such requirements with respect to construction or physical condition as may be specified by the Secretary of State;

(i) to ensure that there is compliance with such requirements with respect to the provision or condition of services and amenities to or within the building as are so specified;

(j) any other purpose for the time being specified for the purposes of this section by order of the Secretary of State.

(2) If in the opinion of the local housing authority the relevant works are more or less extensive than is necessary to achieve any of the purposes set out in subsection (1), they may, with the consent of the applicant, treat the application as varied so that the relevant works are limited to or, as the case may be, include such works as seem to the authority to be necessary for that purpose.

(3) In exercise of the powers conferred by paragraphs (h) and (i) of subsection (1) the Secretary of State may specify requirements generally or for particular cases, and may specify different requirements for different areas.

18   Common parts grants: approval of application

(1) The local housing authority may approve an application for a common parts grant if they think fit, subject to the following provisions.

(2) The authority shall not approve an application for a common parts grant unless they are satisfied that the works are necessary for one or more of the purposes set out in section 17(1).

(3) An authority proposing to approve an application for a common parts grant shall consider whether the building to which the application relates meets the requirements mentioned in paragraphs (a) to (e) of section 604(2) of the Housing Act 1985.

(4) If it appears to the authority that the building does not meet those requirements, they shall not approve the application unless they are satisfied—

(a) that on completion of the relevant works, together with any other works proposed to be carried out, the building will meet those requirements,

(b) that there are satisfactory financial and other arrangements for carrying out those works, and

(c) that the carrying out of the works is the most satisfactory course of action.

(5) In considering whether to approve an application for a common parts grant the local housing authority shall have regard to the expected life of the building (taking account, where appropriate, of the effect of carrying out the works).

Disabled facilities grants

19   Disabled facilities grants: owner's and tenant's applications

(1) A local housing authority shall not entertain an application for a disabled facilities grant unless they are satisfied—

  (a) that the applicant has, or proposes to acquire, an owner's interest in every parcel of land on which the relevant works are to be carried out, or

  (b) that the applicant is a tenant (alone or jointly with others)—

       (i) in the case of an application in respect of works to a dwelling, of the dwelling, or

       (ii) in the case of a common parts application, of a flat in the building,

   and, in either case, does not have or propose to acquire such an owner's interest as is mentioned in paragraph (a).

(2) References in this Chapter to an "owner's application" or a "tenant's application", in relation to a disabled facilities grant, shall be construed accordingly.

(3) In accordance with directions given by the Secretary of State, a local housing authority may treat the condition in subsection (1)(a) as met by a person who has, or proposes to acquire, an owner's interest in only part of the land concerned.

(4) In this Chapter, in relation to an application for a disabled facilities grant—

       "qualifying owner's interest" means an owner's interest meeting the condition in subsection (1)(a) or treated by virtue of subsection (3) as meeting that condition; and

       "qualifying tenant" means a tenant who meets the conditions in subsection (1)(b).

(5) In this Chapter"tenant", in relation to a disabled facilities grant, includes—

  (a) a secure tenant, introductory tenant or statutory tenant,

  (b) a protected occupier under the Rent (Agriculture) Act 1976 or a person in occupation under an assured agricultural occupancy within the meaning of Part I of the Housing Act 1988,

  (c) an employee (whether full-time or part-time) who occupies the dwelling or flat concerned for the better performance of his duties, and

  (d) a person having a licence to occupy the dwelling or flat concerned which satisfies such conditions as may be specified by order of the Secretary of State;

and other expressions relating to tenancies, in the context of an application for disabled facilities grant, shall be construed accordingly.

20   Disabled facilities grants: the disabled occupant

In this Chapter the"disabled occupant", in relation to an application for disabled facilities grant, means the disabled person for whose benefit it is proposed to carry out any of the relevant works.

21   Disabled facilities grants: certificate required in case of owner's application

(1) A local housing authority shall not entertain an owner's application for a disabled facilities grant unless it is accompanied by an owner's certificate in respect of the

dwelling to which the application relates or, in the case of a common parts application, in respect of each flat in the building occupied or proposed to be occupied by a disabled occupant.

(2) An"owner's certificate", for the purposes of an application for a disabled facilities grant, certifies that the applicant—

    (a)   has or proposes to acquire a qualifying owner's interest, and

    (b)   intends that the disabled occupant will live in the dwelling or flat as his only or main residence throughout the grant condition period or for such shorter period as his health and other relevant circumstances permit.

22   Disabled facilities grants: certificates required in case of tenant's application

(1) A local housing authority shall not entertain a tenant's application for a disabled facilities grant unless it is accompanied by a tenant's certificate.

(2) A"tenant's certificate", for the purposes of an application for a disabled facilities grant, certifies—

    (a)   that the application is a tenant's application, and

    (b)   that the applicant intends that he (if he is the disabled occupant) or the disabled occupant will live in the dwelling or flat as his only or main residence throughout the grant condition period or for such shorter period as his health and other relevant circumstances permit.

(3) Except where the authority consider it unreasonable in the circumstances to require such a certificate, they shall not entertain a tenant's application for a disabled facilities grant unless it is also accompanied by an owner's certificate from the person who at the time of the application is the landlord under the tenancy.

23   Disabled facilities grants: purposes for which grant must or may be given

(1) The purposes for which an application for a disabled facilities grant must be approved, subject to the provisions of this Chapter, are the following—

    (a)   facilitating access by the disabled occupant to and from the dwelling or the building in which the dwelling or, as the case may be, flat is situated;

    (b)   making the dwelling or building safe for the disabled occupant and other persons residing with him;

    (c)   facilitating access by the disabled occupant to a room used or usable as the principal family room;

    (d)   facilitating access by the disabled occupant to, or providing for the disabled occupant, a room used or usable for sleeping;

    (e)   facilitating access by the disabled occupant to, or providing for the disabled occupant, a room in which there is a lavatory, or facilitating the use by the disabled occupant of such a facility;

    (f)   facilitating access by the disabled occupant to, or providing for the disabled occupant, a room in which there is a bath or shower (or both), or facilitating the use by the disabled occupant of such a facility;

    (g)   facilitating access by the disabled occupant to, or providing for the disabled occupant, a room in which there is a washhand basin, or facilitating the use by the disabled occupant of such a facility;

    (h)   facilitating the preparation and cooking of food by the disabled occupant;

(i)   improving any heating system in the dwelling to meet the needs of the disabled occupant or, if there is no existing heating system in the dwelling or any such system is unsuitable for use by the disabled occupant, providing a heating system suitable to meet his needs;

(j)   facilitating the use by the disabled occupant of a source of power, light or heat by altering the position of one or more means of access to or control of that source or by providing additional means of control;

(k)   facilitating access and movement by the disabled occupant around the dwelling in order to enable him to care for a person who is normally resident in the dwelling and is in need of such care;

(l)   such other purposes as may be specified by order of the Secretary of State.

(2) An application for a disabled facilities grant may be approved, subject to the provisions of this Chapter, for the purpose of making the dwelling or building suitable for the accommodation, welfare or employment of the disabled occupant in any other respect.

(3) If in the opinion of the local housing authority the relevant works are more or less extensive than is necessary to achieve any of the purposes set out in subsection (1) or the purpose mentioned in subsection (2), they may, with the consent of the applicant, treat the application as varied so that the relevant works are limited to or, as the case may be, include such works as seem to the authority to be necessary for that purpose.

24   Disabled facilities grants: approval of application

(1) The local housing authority—

(a)   shall approve an application for a disabled facilities grant for purposes within section 23(1), and

(b)   may if they think fit approve an application for a disabled facilities grant not for a purpose within that provision but for the purpose specified in section 23(2),

subject to the following provisions.

(2) Where an authority entertain an owner's application for a disabled facilities grant made by a person who proposes to acquire a qualifying owner's interest, they shall not approve the application until they are satisfied that he has done so.

(3) A local housing authority shall not approve an application for a disabled facilities grant unless they are satisfied—

(a)   that the relevant works are necessary and appropriate to meet the needs of the disabled occupant, and

(b)   that it is reasonable and practicable to carry out the relevant works having regard to the age and condition of the dwelling or building.

In considering the matters mentioned in paragraph (a) a local housing authority which is not itself a social services authority shall consult the social services authority.

(4) An authority proposing to approve an application for a disabled facilities grant shall consider—

(a)   in the case of an application in respect of works to a dwelling, whether the dwelling is fit for human habitation;

(b)   in the case of a common parts application, whether the building meets the requirements in section 604(2) of the Housing Act 1985.

and the authority shall take that into account in deciding whether it is reasonable and practicable to carry out the relevant works.

(5) A local housing authority shall not approve a common parts application for a disabled facilities grant unless they are satisfied that the applicant has a power or is under a duty to carry out the relevant works.

HMO grants

25   HMO grants: the interest of the applicant in the property

(1) A local housing authority shall not entertain an application for an HMO grant unless they are satisfied that the applicant has or proposes to acquire an owner's interest in every parcel of land on which the relevant works are to be carried out.

(2) In accordance with directions given by the Secretary of State, a local housing authority may treat the condition in subsection (1) as fulfilled by a person who has, or proposes to acquire, an owner's interest in only part of the land concerned.

(3) References in this Chapter to "a qualifying owner's interest", in relation to an application for an HMO grant, are to an owner's interest meeting the condition in subsection (1) or treated by virtue of subsection (2) as meeting that condition.

26   HMO grants: certificate required to accompany application

(1) A local housing authority shall not entertain an application for an HMO grant unless it is accompanied by a certificate of future occupation.

(2) A "certificate of future occupation" certifies that the applicant—
  (a) has or proposes to acquire a qualifying owner's interest in the house, and
  (b) intends that throughout the grant condition period the house or a part of it (specified in the certificate) will be residentially occupied, or available for residential occupation, under tenancies or licences by persons who are not connected with the owner for the time being of the house.

In paragraph (b) "residential occupation" does not include occupation for a holiday, and "tenancies" does not include a long tenancy.

27   HMO grants: purposes for which grant may be given

(1) The purposes for which an application for an HMO grant (other than a conversion application) may be approved are—
  (a) to comply with a notice under section 189 of the Housing Act 1985 (repair notice in respect of unfit premises) or otherwise to render the house fit for human habitation;
  (b) to comply with a notice under section 190 of that Act (repair notice in respect of premises not unfit but in need of substantial repair) or otherwise to put the building in reasonable repair;
  (c) to comply with a notice under section 352 of that Act (notice requiring works to render premises fit for the number of occupants) or otherwise to enable the house to meet one or more of the requirements in subsection (1A) of that section;

    (d)   to provide adequate thermal insulation;

    (e)   to provide adequate facilities for space heating;

    (f)   to provide satisfactory internal arrangements;

    (g)   to provide means of escape in case of fire or other fire precautions, not being precautions required under or by virtue of any enactment (whenever passed);

    (h)   to ensure that the house complies with such requirements with respect to construction or physical condition as may be specified by the Secretary of State;

    (i)   to ensure that there is compliance with such requirements with respect to the provision or condition of services and amenities to or within the house as are so specified;

    (j)   any other purpose for the time being specified for the purposes of this section by order of the Secretary of State.

(2) The purpose for which a conversion application may be approved is to provide a house in multiple occupation by the conversion of a house or other building.

(3) If in the opinion of the authority the relevant works are more or less extensive than is necessary to achieve any of the purposes set out in subsection (1) or (2), they may, with the consent of the applicant, treat the application as varied so that the relevant works are limited to or, as the case may be, include such works as seem to the authority to be necessary for that purpose.

(4) In exercise of the powers conferred by paragraphs (h) and (i) of subsection (1) the Secretary of State may specify requirements generally or for particular cases, and may specify different requirements for different areas.

28   HMO grants: approval of application

(1) The local housing authority may approve an application for an HMO grant if they think fit, subject to the following provisions.

(2) The authority shall not approve an application for an HMO grant unless they are satisfied that the works are necessary for one or more of the purposes set out in section 27(1) or (2).

(3) Where an authority entertain an application for an HMO grant made by a person who proposes to acquire a qualifying owner's interest, they shall not approve the application until they are satisfied that he has done so.

(4) An authority proposing to approve an application for an HMO grant shall consider whether the house to which the application relates is fit for human habitation and meets the requirements in section 352(1A) of the Housing Act 1985.

(5) If it appears to the authority that the house is not fit for human habitation or does not meet those requirements, they shall not approve the application unless they are satisfied—

    (a)   that on completion of the relevant works, together with any other works proposed to be carried out, the house will be fit for human habitation and meet those requirements,

    (b)   that there are satisfactory financial and other arrangements for carrying out those works, and

    (c)   that the carrying out of the works is the most satisfactory course of action.

Status: *This is the original version (as it was originally enacted).*

(6) In considering whether to approve an application for an HMO grant the local housing authority shall have regard to the expected life of the house (taking account, where appropriate, of the effect of carrying out the works).

Restrictions on grant aid

29   Restriction on grants for works already begun

(1) Subject as follows, a local housing authority shall not approve an application for a grant if the relevant works have been begun before the application is approved.

(2) Where the relevant works have been begun but have not been completed, the authority may approve the application for a grant if they are satisfied that there were good reasons for beginning the works before the application was approved.

(3) Where an authority decide to approve an application in accordance with subsection (2), they may, with the consent of the applicant, treat the application as varied so that the relevant works do not include any that are completed.

But in determining for the purposes of the application the physical condition of the dwelling, common parts or house or other building concerned, they shall consider the condition of the premises at the date of the application.

(4) Subject as follows, a local housing authority shall not approve an application for a grant if the relevant works have been completed.

(5) Nothing in this section applies to an application for a grant in respect of works necessary—
  (a)   to render a dwelling fit for human habitation or to comply with a notice under section 189 or 190 of the Housing Act 1985 (repair notices), or
  (b)   to enable a house in multiple occupation to meet one or more of the requirements in section 352(1A) of that Act (fitness for the number of occupants) or to comply with a notice under that section.

(6) If the local housing authority consider that the relevant works include works in addition to those necessary for the purposes mentioned in subsection (5)(a) or (b), they shall treat the application as an application to which this section applies so far as it relates to those additional works.

30   Means testing in case of application by owner-occupier or tenant

(1) This section applies—
  (a)   to an application for a renovation grant which is—
      (i) an owner's application accompanied by an owner-occupation certificate, or
      (ii) a tenant's application; and
  (b)   to any application for a disabled facilities grant.

(2) An owner's application for a renovation grant shall be treated as falling within this section if it is a conversion application for the provision of two or more dwellings and any of the certificates accompanying the application is an owner-occupation certificate.

(3) If in the case of an application for a renovation grant to which this section applies the financial resources of the applicant exceed the applicable amount, the amount of any grant which may be paid shall, in accordance with regulations, be reduced from what it would otherwise have been.

(4) If in the case of an application for a disabled facilities grant the financial resources of any person of a description specified by regulations exceed the applicable amount, the amount of any grant which may be paid shall, in accordance with regulations, be reduced from what it would otherwise have been.

(5) Provision may be made by regulations—

    (a)   for the determination of the amount which is to be taken to be the financial resources of any person,

    (b)   for the determination of the applicable amount referred to in subsection (3) or (4), and

    (c)   as to circumstances in which the financial resources of a person are to be assumed (by reason of his receiving a prescribed benefit or otherwise) not to exceed the applicable amount.

(6) Regulations may, in particular—

    (a)   make provision for account to be taken of the income, assets, needs and outgoings not only of the person himself but also of his spouse, any person living with him or intending to live with him and any person on whom he is dependent or who is dependent on him;

    (b)   make provision for amounts specified in or determined under the regulations to be taken into account for particular purposes.

(7) Regulations may apply for the purposes of this section, subject to such modifications as may be prescribed, any other statutory means-testing regime as it has effect from time to time.

(8) Regulations may make provision requiring any information or evidence needed for the determination of any matter under this section to be furnished by such person as may be prescribed.

(9) In this section "regulations" means regulations made by the Secretary of State with the consent of the Treasury.

31   Determination of amount of grant in case of landlord's application

(1) This section applies to—

    (a)   an owner's application for a renovation grant which is accompanied by a certificate of intended letting (not being an application which falls within section 30: see subsection (2) of that section),

    (b)   a landlord's application for a common parts grant,

    (c)   a landlord's application for a disabled facilities grant, and

    (d)   any application for an HMO grant.

(2) The reference in subsection (1)(c) to a landlord's application for a disabled facilities grant is to an owner's application in respect of works to a dwelling which is or is intended to be let, or to the common parts of a building in which a flat is or is intended to be let.

(3) The amount of the grant (if any) shall be determined by the local housing authority, having regard to—

    (a) the extent to which the landlord is able to charge a higher rent for the premises because of the works, and

    (b) such other matters as the Secretary of State may direct.

(4) The authority may, if they think it appropriate, seek and act upon the advice of rent officers as to any matter.

(5) The Secretary of State may by regulations make provision requiring any information or evidence needed for the determination of any matter under this section to be furnished by such person as may be prescribed.

32   Apportionment in case of tenants' application for common parts grant

(1) This section applies where a local housing authority approve a tenants' application for a common parts grant.

(2) The local housing authority shall decide how much of the cost of the relevant works is attributable to the applicants ("the attributable cost").

(3) For the purposes of this section the attributable cost is an amount equal to the following proportion of the cost of the relevant works—

    (a) if it can be ascertained, the proportion that the aggregate of the respective liabilities of each of the applicants to carry out or contribute to the carrying out of the relevant works bears to the aggregate of all such liabilities on the part of all persons (including the applicants) so liable; or

    (b) if the proportion mentioned in paragraph (a) cannot be ascertained, the proportion that the number of applicants bears to the number of persons (including the applicants) liable to carry out or contribute to the carrying out of works to the building.

(4) The local housing authority shall then apportion the attributable cost to each of the applicants—

    (a) in a case where the attributable cost is calculated by reference to the proportion mentioned in subsection (3)(a), according to the proportion that his liabilities to carry out or contribute to the carrying out of the relevant works bears to the aggregate of the applicants' liabilities mentioned in that paragraph; or

    (b) in a case where the attributable cost is calculated by reference to the proportion mentioned in subsection (3)(b), equally.

(5) The amount of the grant payable shall be the aggregate of the grants that would be payable to each of the applicants under section 30 or, in the case of a participating landlord, under section 31 if each of the applicants was an individual applicant in respect of his portion of the attributable cost.

(6) Where the interest of an occupying tenant is held jointly by two or more persons, those persons shall be regarded as a single person for the purposes of this section.

33   Power to specify maximum amount of grant

(1) The Secretary of State may, if he thinks fit, by order specify a maximum amount or a formula for calculating a maximum amount of grant which a local housing authority may pay in respect of an application for a grant.

(2) An order under this section may make different provision for different types of grant, or for the same type of grant in different circumstances.

(3) In relation to an application for a grant in respect of works for any of the purposes in section 23(1) (mandatory disabled facilities grant), the order may—

    (a)   provide for a maximum amount of grant to be paid if the application is approved, and

    (b)   authorise the local housing authority, if they think fit, to pay a further amount in excess of that maximum but subject to such other maximum (if any) as may be specified in or determined in accordance with the order.

(4) An authority may not, except as mentioned in subsection (3), pay an amount of grant in excess of a specified maximum amount.

Decision and notification

34   Decision and notification

(1) A local housing authority shall by notice in writing notify an applicant for a grant as soon as reasonably practicable, and, in any event, not later than six months after the date of the application concerned, whether the application is approved or refused.

(2) Where an authority decide to approve an application for a grant, they shall determine—

    (a)   which of the relevant works are eligible for grant (in this Chapter referred to as "the eligible works"),

    (b)   the amount of the expenses which in their opinion are properly to be incurred in the execution of the eligible works,

    (c)   the amount of the costs which in their opinion have been properly incurred, or are properly to be incurred, with respect to preliminary or ancillary services and charges, and

    (d)   the amount of grant they have decided to pay, taking into account all the relevant provisions of this Chapter.

The total of the amounts referred to in paragraphs (b) and (c) is referred to in this Chapter as "the estimated expense".

(3) If the authority notify the applicant under subsection (1) that the application is approved, they shall specify in the notice—

    (a)   the eligible works,

    (b)   the amounts referred to in subsection (2)(b) and (c), and how those amounts have been calculated, and

    (c)   the amount of the grant.

(4) If the authority notify the applicant under subsection (1) that the application is refused, they shall at the same time notify him of the reasons for the refusal.

(5) If after an application for a grant has been approved the authority are satisfied that owing to circumstances beyond the control of the applicant—

(a) the eligible works cannot be, or could not have been, carried out on the basis of the amount of expenses referred to in subsection (2)(b),

(b) the amount of the costs which have been or are to be incurred as mentioned in subsection (2)(c) has increased, or

(c) the eligible works cannot be, or could not have been, carried out without carrying out additional works which could not have been reasonably foreseen at the time the application was made,

the authority may re-determine the estimated expense and the amount of the grant.

(6) Where an application for a grant is approved, the local housing authority may not impose any condition in relation to the approval or payment of the grant, except—

(a) as provided by the following provisions of this Chapter, or

(b) with the consent of the Secretary of State;

and this applies whether the condition purports to operate as a condition, a personal covenant or otherwise.

Payment of grants

35 Payment of grants: general

(1) Where the local housing authority have approved an application for a grant, they shall pay the grant, subject to the following provisions of this Chapter.

(2) The grant may be paid—

(a) in whole after the completion of the eligible works, or

(b) in part by instalments as the works progress and the balance after completion of the works.

(3) Where a grant is paid by instalments, the aggregate of the instalments paid before the completion of the eligible works shall not at any time exceed nine-tenths of the amount of the grant.

36 Delayed payment of mandatory grant

(1) Where the local housing authority are obliged to approve an application for a grant under section 24(1)(a) (mandatory disabled facilities grant), they may do so on terms that payment of the grant, or part of it, will not be made before a date specified in the notification of their decision on the application.

(2) That date shall not be more than twelve months, or such other period as may be specified by order of the Secretary of State, after the date of the application.

37 Payment of grants: conditions as to carrying out of the works

(1) It is a condition of payment of every grant that the eligible works are carried out within twelve months from—

(a) the date of approval of the application concerned, or

(b) where section 36 applies (delayed payment of mandatory grant), the date specified in the notification of the authority's decision,

or, in either case, such further period as the local housing authority may allow.

(2) The authority may, in particular, allow further time where they are satisfied that the eligible works cannot be, or could not have been, carried out without carrying out other works which could not have been reasonably foreseen at the time the application was made.

(3) In approving an application for a grant a local housing authority may require as a condition of payment of the grant that the eligible works are carried out in accordance with such specifications as they determine.

(4) The payment of a grant, or part of a grant, is conditional upon—

    (a) the eligible works or the corresponding part of the works being executed to the satisfaction of the authority, and

    (b) the authority being provided with an acceptable invoice, demand or receipt for payment for the works and any preliminary or ancillary services or charges in respect of which the grant or part of the grant is to be paid.

For this purpose an invoice, demand or receipt is acceptable if it satisfies the authority and is not given by the applicant or a member of his family.

38   Payment of grants: conditions as to contractors employed

(1) It is a condition of payment of every grant, unless the local housing authority direct otherwise in any particular case, that the eligible works are carried out by the contractor whose estimate accompanied the application or, where two or more estimates were submitted, by one of those contractors.

(2) The Secretary of State may by regulations make provision as to the establishing and maintaining by local housing authorities of lists of contractors approved by them for the purpose of carrying out grant-aided works.

(3) The regulations may provide that it shall be a condition of payment of every grant by a local housing authority by whom such a list is maintained that, except in such cases as may be prescribed and unless the local housing authority direct otherwise in any particular case, the eligible works are carried out by a contractor who is on the authority's list of approved contractors.

39   Payment of grant to contractor

(1) The local housing authority may pay a grant or part of a grant—

    (a) by payment direct to the contractor, or

    (b) by delivering to the applicant an instrument of payment in a form made payable to the contractor.

They shall not do so unless the applicant was informed before the grant application was approved that this would or might be the method of payment.

(2) Where an amount of grant is payable, but the works in question have not been executed to the satisfaction of the applicant, the local housing authority may at the applicant's request and if they consider it appropriate to do so withhold payment from the contractor.

If they do so, they may make the payment to the applicant instead.

40   Applicant ceasing to be entitled before payment of grant

(1) This section applies where an application for a grant is approved but before the certified date the applicant ceases to be a person entitled to a grant of that description.

In the case of a joint application this section does not apply unless all the applicants cease to be so entitled.

(2) Where this section applies—

    (a)   in the case of a renovation grant, disabled facilities grant or HMO grant, no grant shall be paid or, as the case may be, no further instalments shall be paid, and

    (b)   in the case of a common parts grant approved on a landlord's application, the local housing authority may refuse to pay the grant or any further instalment,

and the authority may demand that any instalment of the grant which has been paid be repaid forthwith, together with interest from the date on which it was paid until repayment, at such reasonable rate as the authority may determine.

(3) For the purposes of this section an applicant ceases to be a person entitled to a renovation grant—

    (a)   in the case of an owner's application—

        (i)   if he ceases to have a qualifying owner's interest, or

        (ii)   if he ceases to have the intention specified in the owner-occupation certificate or certificate of intended letting which accompanied the application;

    (b)   in the case of a tenant's application—

        (i)   if he ceases to be a qualifying tenant of the dwelling, or

        (ii)   if the application was accompanied by a certificate of intended letting and the landlord ceases to have the intention specified in the certificate; or

    (c)   if the application was approved under section 13(5) (approval of grant in respect of works to unfit premises) and the authority cease to be satisfied of the matters mentioned in that provision.

(4) For the purposes of this section an applicant ceases to be a person entitled to a disabled facilities grant—

    (a)   in the case of an owner's application—

        (i)   if he ceases to have a qualifying owner's interest, or

        (ii)   if he ceases to have the intention specified in the owner's certificate which accompanied the application;

    (b)   in the case of a tenant's application—

        (i)   if he ceases to be a qualifying tenant of the dwelling, or

        (ii)   if the application was accompanied by an owner's certificate and the landlord ceases to have the intention specified in the certificate.

But if the case falls within section 41 (change of circumstances affecting disabled occupant), the authority shall act under that section.

(5) For the purposes of this section an applicant ceases to be a person entitled to an HMO grant—

    (a)   if he ceases to have a qualifying owner's interest in the house;

   (b)   if he ceases to have the intention specified in the certificate of future occupation which accompanied the application; or

   (c)   if the application was approved under section 28(5) (approval of grant in respect of works to unfit premises) and the authority cease to be satisfied of the matters mentioned in that provision.

(6) For the purposes of this section an applicant whose application is a landlord's application for a common parts grant ceases to be a person entitled to a grant—

   (a)   if he ceases to have an owner's interest in the building;

   (b)   if he ceases to have a duty or power to carry out the relevant works; or

   (c)   if the application was approved under section 18(4) (approval of grant in respect of works to unfit premises) and the authority cease to be satisfied of the matters mentioned in that provision.

(7) This section has effect subject to section 56 (provisions relating to death of applicant).

41   Change of circumstances affecting disabled occupant

(1) This section applies where an application for a disabled facilities grant has been approved and before the certified date—

   (a)   the works cease to be necessary or appropriate to meet the needs of the disabled occupant, or

   (b)   the disabled occupant ceases to occupy the dwelling or flat concerned or it ceases to be the intention that he should occupy it, or

   (c)   the disabled occupant dies.

Where the application related to more than one disabled occupant, this section applies if any of paragraphs (a) to (c) applies in relation to any of them.

(2) This section applies whether or not the disabled occupant (or any of them) is the applicant (or one of them).

(3) Where this section applies the local housing authority may take such action as appears to them appropriate and may decide—

   (a)   that no grant shall be paid or, as the case may be, no further instalments shall be paid,

   (b)   that the relevant works or some of them should be completed and the grant or an appropriate proportion of it paid, or

   (c)   that the application should be redetermined in the light of the new circumstances.

(4) In making their decision the authority shall have regard to all the circumstances of the case.

(5) If the authority decide that no grant shall be paid or that no further instalments shall be paid, they may demand that any instalment of the grant which has been paid be repaid forthwith, together with interest from the date on which it was paid until repayment, at such reasonable rate as the authority may determine.

42   Cases in which grants may be re-calculated, withheld or repaid

(1) This section applies where an application for a grant has been approved by the local housing authority and—

- (a) the authority ascertain that the amount was determined under section 30 or 31 on the basis of inaccurate or incomplete information and exceeds that to which the applicant was entitled;
- (b) the authority ascertain that without their knowledge the eligible works were started before the application was approved;
- (c) the eligible works are not completed to the satisfaction of the authority within the period specified under section 37(1), or such extended period as they may allow under that provision;
- (d) the authority ascertain that the aggregate of the cost of completing the eligible works and the costs incurred with respect to preliminary or ancillary services and charges, is or is likely to be lower than the estimated expense; or
- (e) the authority ascertain that without their knowledge the eligible works were carried out otherwise than as required by section 38 (conditions as to contractors employed).

(2) Where this section applies, the authority may—

- (a) refuse to pay the grant or any further instalment of grant which remains to be paid, or
- (b) make a reduction in the grant which, in a case falling within subsection (1)(d), is to be a reduction proportionate to the reduction in the estimated expense;

and they may demand repayment by the applicant forthwith, in whole or part, of the grant or any instalment of the grant paid, together with interest at such reasonable rate as the authority may determine from the date of payment until repayment.

43 Repayment where applicant not entitled to grant

(1) This section applies where an application for a grant is approved but it subsequently appears to the local housing authority that the applicant (or, in the case of a joint application, any of the applicants) was not, at the time the application was approved, entitled to a grant of that description.

(2) Where this section applies—

- (a) in the case of a renovation grant, disabled facilities grant or HMO grant, no grant shall be paid or, as the case may be, no further instalments shall be paid, and
- (b) in the case of a common parts grant approved on a landlord's application, the local housing authority may refuse to pay the grant or any further instalment,

and the authority may demand that any grant which has been paid be repaid forthwith, together with interest from the date on which it was paid until repayment, at such reasonable rate as the authority may determine.

(3) For the purposes of this section an applicant is not entitled to a renovation grant—

- (a) in the case of an owner's application if—
  - (i) he does not have a qualifying owner's interest, or
  - (ii) he does not have the intention specified in the owner-occupation certificate or certificate of intended letting which accompanied the application; or
- (b) in the case of a tenant's application if—
  - (i) he is not a qualifying tenant of the dwelling, or

(ii) if the application was accompanied by a certificate of intended letting and the landlord does not have the intention specified in the certificate.

(4) For the purposes of this section an applicant is not entitled to a disabled facilities grant—

   (a)   in the case of an owner's application—

      (i) if he does not have a qualifying owner's interest, or

      (ii) if he does not have the intention specified in the owner's certificate which accompanied the application; or

   (b)   in the case of a tenant's application—

      (i) if he is not a qualifying tenant of the dwelling, or

      (ii) if the application was accompanied by an owner's certificate and the landlord does not have the intention specified in the certificate.

(5) For the purposes of this section an applicant is not entitled to an HMO grant—

   (a)   if he does not have a qualifying owner's interest in the house; or

   (b)   if he does not have the intention specified in the certificate of future occupation which accompanied the application.

(6) For the purposes of this section an applicant whose application is a landlord's application for a common parts grant is not entitled to a grant—

   (a)   if he does not have an owner's interest in the building; or

   (b)   if he does not have a duty or power to carry out the relevant works.

Grant conditions and repayment

44   Grant conditions: introductory

(1) The following sections have effect with respect to the conditions to be observed where an application for a grant has been approved by a local housing authority.

In this Chapter a "grant condition" means a condition having effect in accordance with any of those sections.

(2) Except as otherwise provided—

   (a)   the grant conditions as to repayment on disposal (sections 45 to 47) have effect from the date on which the application is approved until the end of the grant condition period;

   (b)   the grant conditions as to occupation (sections 48 to 50) have effect from the certified date until the end of the grant condition period; and

   (c)   a grant condition imposed under section 52 (power to impose other conditions with consent of Secretary of State) has effect for such period as may be specified in, or in accordance with, the Secretary of State's consent.

(3) In this Chapter—

   (a)   the "grant condition period" means the period of five years, or such other period as the Secretary of State may by order specify or as may be imposed by the local housing authority with the consent of the Secretary of State, beginning with the certified date; and

(b) the "certified date" means the date certified by the local housing authority as the date on which the execution of the eligible works is completed to their satisfaction.

(4) A local housing authority may not impose any condition requiring a grant to be repaid except in accordance with the following sections.

This applies whether the condition purports to operate as a condition of the grant, as a personal covenant or otherwise.

45   Condition for repayment on disposal: renovation grants

(1) It is a condition of a renovation grant that if an owner of the premises to which the application relates makes a relevant disposal (other than an exempt disposal)—

    (a) of the whole or part of the premises to which the application relates,

    (b) after any instalment of grant has been paid, and

    (c) before the certified date,

he shall repay to the local housing authority on demand the amount of grant that has been paid.

(2) It is a condition of a renovation grant that if an owner of the dwelling to which the application relates or, in the case of a conversion application, any dwelling provided by the relevant works, makes a relevant disposal (other than an exempt disposal)—

    (a) of the whole or part of the dwelling,

    (b) on or after the certified date, and

    (c) before the end of the grant condition period,

he shall repay to the local housing authority on demand the amount of grant that has been paid.

In the case of a conversion application the grant shall be treated for this purpose as apportioned equally between the dwellings provided.

(3) A condition under this section is a local land charge and is binding on any person who is for the time being an owner of the premises concerned.

(4) Where the authority have the right to demand repayment of an amount as mentioned in subsection (1) or (2), they may—

    (a) if the case falls within subsection (5), or

    (b) in any other case, with the consent of the Secretary of State,

determine not to demand payment or to demand a lesser amount.

(5) The cases referred to in subsection (4)(a) are where the authority are satisfied that the owner of the dwelling—

    (a) is elderly or infirm and is making the disposal with the intention—

        (i) of going to live in a hospital, hospice, sheltered housing, residential care home or similar institution as his only or main residence, or

        (ii) of moving to somewhere where care will be provided by any person; or

    (b) is making the disposal with the intention of going to live with and care for an elderly or infirm member of his family or his partner's family.

(6) Any condition under this section shall cease to be in force with respect to any premises if there is a relevant disposal of the premises that is an exempt disposal, other than—

    (a) a disposal within section 54(1)(a) (disposal to associates of person making disposal), or

    (b) a disposal within section 54(1)(b) (vesting under will or on intestacy).

46   Condition for repayment on disposal: common parts grants

(1) It is a condition of a common parts grant approved on a landlord's application that if the applicant makes a relevant disposal (other than an exempt disposal)—

    (a) of the whole or part of the building,

    (b) after any instalment of grant has been paid, and

    (c) before the certified date,

he shall repay to the local housing authority on demand the amount of grant that has been paid.

(2) It is a condition of a common parts grant approved on a landlord's application that if the applicant makes a relevant disposal (other than an exempt disposal)—

    (a) of the whole or part of the building,

    (b) on or after the certified date, and

    (c) before the end of the grant condition period,

he shall repay to the local housing authority on demand the amount of grant that has been paid.

(3) A condition under this section is a local land charge and is binding on any person who is for the time being a successor in title to the interest in the building by virtue of which the applicant made his application.

(4) Where the authority have the right to demand repayment of an amount as mentioned in subsection (1) or (2), they may, with the consent of the Secretary of State, determine not to demand payment or to demand a lesser amount.

(5) Any condition under this section shall cease to be in force with respect to any premises if there is a relevant disposal of the premises that is an exempt disposal.

47   Condition for repayment on disposal: HMO grants

(1) It is a condition of an HMO grant that if an owner of the house makes a relevant disposal (other than an exempt disposal)—

    (a) of the whole or part of the house,

    (b) after any instalment of grant has been paid, and

    (c) before the certified date,

he shall repay to the local housing authority on demand the amount of grant that has been paid.

(2) It is a condition of an HMO grant that if an owner of the house makes a relevant disposal (other than an exempt disposal)—

    (a) of the whole or part of the house,

    (b) on or after the certified date, and

    (c) before the end of the grant condition period,

he shall repay to the local housing authority on demand the amount of grant that has been paid.

(3) A condition under this section is a local land charge and is binding on any person (other than a local housing authority or registered social landlord) who is for the time being an owner of the house.

(4) Where the authority have the right to demand repayment of an amount as mentioned in subsection (1) or (2), they may, with the consent of the Secretary of State, determine not to demand payment or to demand a lesser amount.

(5) Any condition under this section shall cease to be in force with respect to any premises if there is a relevant disposal of the premises that is an exempt disposal.

48   Condition as to owner-occupation: renovation grants

(1) Where an application for a renovation grant was accompanied by an owner-occupation certificate in respect of any dwelling (see section 8(2)), it is a condition of the grant that throughout the grant condition period the dwelling is occupied in accordance with the intention stated in the certificate.

(2) It is also a condition of the grant that if at any time when that condition is in force the authority serve notice on the owner of the dwelling requiring him to do so, he will within the period of 21 days beginning with the date on which the notice was served furnish to the authority a statement showing how that condition is being fulfilled.

(3) A condition under this section is a local land charge and is binding on any person who is for the time being an owner of the dwelling.

(4) In the event of a breach of a condition under this section, the owner for the time being of the dwelling shall on demand repay to the local housing authority the amount of the grant, together with compound interest on that amount as from the certified date, calculated at such reasonable rate as the authority may determine and with yearly rests.

(5) The local housing authority may determine not to make such a demand or to demand a lesser amount.

(6) Any condition under this section shall cease to be in force with respect to the dwelling if there is a relevant disposal of the dwelling that is an exempt disposal, other than—

    (a)   a disposal within section 54(1)(a) (disposal to associates of person making disposal), or

    (b)   a disposal within section 54(1)(b) (vesting under will or on intestacy).

49   Condition as to availability for letting: renovation grants

(1) Where an application for a renovation grant was accompanied by a certificate of intended letting in respect of any dwelling (see section 8(3)), it is a condition of the grant that throughout the grant condition period the dwelling is let or available for letting in accordance with the intention stated in the certificate.

(2) It is also a condition of the grant that if at any time within the grant condition period the local housing authority by whom the grant was paid serve notice on the owner of the dwelling requiring him to do so, he will within the period of 21 days beginning on the date on which the notice was served furnish to the authority a statement showing how the condition in subsection (1) is being fulfilled.

(3) A condition under this section is a local land charge and is binding on any person (other than a local housing authority or registered social landlord) who is for the time being the owner of the dwelling.

(4) In the event of a breach of a condition under this section, the owner for the time being of the dwelling shall on demand repay to the local housing authority the amount of the grant, together with compound interest on that amount as from the certified date, calculated at such reasonable rate as the authority may determine and with yearly rests.

(5) The local housing authority may determine not to make such a demand or to demand a lesser amount.

(6) The terms of any tenancy of the dwelling (or any part of it, or any property including the dwelling or part of it) shall be deemed to include a duty on the part of the tenant, if required to do so by the owner of the dwelling, to furnish him with such information as he may reasonably require to enable him to comply with a notice under subsection (2).

50   Conditions as to occupation: HMO grants

(1) It is a condition of an HMO grant that throughout the grant condition period—

    (a)   the house is occupied or available for residential occupation in accordance with the intention stated in the certificate of future occupation that accompanied the application (see section 26(2)); and

    (b)   that the house is not so occupied as to cause—

        (i) a breach of the duty under section 353A of the Housing Act 1985 (duty to keep premises fit for number of occupants), or

        (ii) a breach of any direction given by the local housing authority under section 354 of that Act (power to limit number of occupants of house).

(2) It is also a condition of the grant that if at any time within the grant condition period the local housing authority by whom the grant was paid serve notice on the owner of the house requiring him to do so, he will within the period of 21 days beginning with the date on which the notice was served furnish to the authority a statement showing how the condition in subsection (1)(a) is being fulfilled.

(3) A condition under this section is a local land charge and is binding on any person (other than a local housing authority or registered social landlord) who is for the time being an owner of the house.

(4) In the event of a breach of a condition under this section, the owner for the time being of the dwelling shall on demand pay to the local housing authority the amount of the grant, together with compound interest on that amount as from the certified date, calculated at such reasonable rate as the authority may determine and with yearly rests.

(5) The local housing authority may determine not to make such a demand or to demand a lesser amount.

(6) The terms of any tenancy of any part of the house shall be deemed to include a duty on the part of the tenant, if required to do so by the owner of the house, to furnish him with such information as he may reasonably require to enable him to comply with a notice under subsection (2).

51   Conditions as to repayment in case of other compensation, &c

(1) Where a local housing authority approve an application for a grant they may, with the consent of the Secretary of State, impose a condition requiring the applicant to take reasonable steps to pursue any relevant claim to which this section applies and to repay the grant, so far as appropriate, out of the proceeds of such a claim.

(2) The claims to which this section applies are—

    (a)   an insurance claim, or a legal claim against another person, in respect of damage to the premises to which the grant relates, or

    (b)   a legal claim for damages in which the cost of the works to premises to which the grant relates is part of the claim;

and a claim is a relevant claim to the extent that works to make good the damage mentioned in paragraph (a), or the cost of which is claimed as mentioned in paragraph (b), are works to which the grant relates.

(3) In the event of a breach of a condition under this section, the applicant shall on demand pay to the local housing authority the amount of the grant so far as relating to any such works, together with compound interest as from such date as may be prescribed by or determined in accordance with the regulations, calculated at such reasonable rate as the authority may determine and with yearly rests.

(4) The local housing authority may determine not to make such a demand or to demand a lesser amount.

52   Power to impose other conditions with consent of Secretary of State

(1) Where a local housing authority approve an application for a grant they may, with the consent of the Secretary of State, impose such conditions as they think fit—

    (a)   relating to things done or omitted before the certified date and requiring the repayment to the local housing authority on demand of any instalments of grant paid, or

    (b)   relating to things done or omitted on or after that date and requiring the payment to the local housing authority on demand of a sum equal to the amount of the grant paid;

and, in either case, that amount may be required to be paid together with compound interest on that amount as from the date of payment, calculated at such reasonable rate as the authority may determine and with yearly rests.

(2) A condition under this section is a local land charge and is binding on—

    (a)   any person who is for the time being an owner of the dwelling, house or building, and

    (b)   such other persons (if any) as the authority may, with the consent of the Secretary of State, specify.

(3) The reference in subsection (2)(a) to the owner of the building shall be construed—

    (a)   in the case of a grant condition imposed on a landlord's application for a common parts grant, as a reference to the applicant or any successor in title to the interest in the building by virtue of which the applicant made his application;

    (b)   in the case of a grant condition imposed on an application for an HMO grant, as excluding a local housing authority or registered social landlord.

(4) Where the authority have the right to demand repayment of an amount as mentioned in subsection (1), they may determine not to demand payment or to demand a lesser amount.

(5) Any conditions imposed under this section are in addition to the conditions provided for by sections 45 to 51.

53   Meaning of relevant disposal

(1) A disposal is a relevant disposal for the purposes of the provisions of this Chapter relating to grant conditions if it is—

    (a)   a conveyance of the freehold or an assignment of the lease, or

    (b)   the grant of a lease (other than a mortgage term) for a term of more than 21 years otherwise than at a rack rent.

(2) For the purposes of subsection (1)(b) it shall be assumed—

    (a)   that any option to renew or extend a lease or sub-lease, whether or not forming part of a series of options, is exercised, and

    (b)   that any option to terminate a lease or sub-lease is not exercised.

(3) The grant of an option enabling a person to call for a relevant disposal shall be treated as such a disposal made to him.

54   Meaning of exempt disposal

(1) A disposal is an exempt disposal for the purposes of the provisions of this Chapter relating to grant conditions if it is a disposal of the whole or part of the premises to which the application relates of any of the following descriptions—

    (a)   a conveyance of the freehold or an assignment of the lease where the person, or each of the persons, to whom it is made is a qualifying person (as defined in subsection (2));

    (b)   a vesting in a person taking under a will or on an intestacy;

    (c)   a disposal in pursuance of any such order as is mentioned in subsection (3);

    (d)   a compulsory disposal (see subsection (4));

    (e)   a disposal of property consisting of land included in the dwelling by virtue of section 184 of the Housing Act 1985 (land let with or used for the purposes of the dwelling-house);

    (f)   a disposal under which the interest of a person entitled to assistance by way of repurchase under Part XVI of that Act (assistance for owners of defective housing) is acquired in accordance with Schedule 20 to that Act;

    (g)   a disposal by way of enfranchisement or lease extension under Part I of the Leasehold Reform Act 1967;

    (h)   a disposal in pursuance of an obligation arising under Chapter I or II of Part I of the Leasehold Reform, Housing and Urban Development Act 1993;

    (i)   a disposal on the exercise of a right of first refusal under Part I of the Landlord and Tenant Act 1987 or in accordance with an acquisition order under Part III of that Act;

    (j)   a disposal on the exercise of—

        (i) the right to buy under Part V of the Housing Act 1985, or

> (ii) the right conferred by section 16 of the Housing Act 1996 (right of tenant of registered social landlord to acquire dwelling);

(k) a conveyance of the freehold or an assignment of the lease where—

> (i) the person making the disposal is aged at least 70,

> (ii) the disposal is to provide an annuity income, and

> (iii) the person concerned is entitled to continue to occupy the premises as his only or main residence;

(l) a disposal of any other description specified by order of the Secretary of State for the purposes of this section.

(2) A person is a qualifying person for the purposes of subsection (1)(a) if—

(a) in the case of an individual, he is—

> (i) the person, or one of the persons, by whom the disposal is made;

> (ii) the spouse, or former spouse, of that person or one of those persons; or

> (iii) a member of the family of that person or one of those persons; or

(b) in the case of a company, it is an associated company of the company by whom the disposal is made.

Section 416 of the Income and Corporation Taxes Act 1988 (meaning of associated company) applies in determining whether a company is an associated company of another for the purposes of paragraph (b).

(3) The orders referred to in subsection (1)(c) are orders under—

(a) section 24 or 24A of the Matrimonial Causes Act 1973 (property adjustment orders or orders for the sale of property in connection with matrimonial proceedings);

(b) section 2 of the Inheritance (Provision for Family and Dependants) Act 1975 (orders as to financial provision to be made from estate);

(c) section 17 of the Matrimonial and Family Proceedings Act 1984 (property adjustment orders or orders for the sale of property after overseas divorce, etc.); or

(d) paragraph 1 of Schedule 1 to the Children Act 1989 (orders for financial relief against parents).

(4) For the purposes of subsection (1)(d) a compulsory disposal is a disposal of property which is acquired compulsorily, or is acquired by a person who has made or would have made, or for whom another person has made or would have made, a compulsory purchase order authorising its compulsory purchase for the purposes for which it is acquired.

(5) The grant of an option enabling a person to call for an exempt disposal shall be treated as such a disposal made to him.

55 Cessation of conditions on repayment of grant, &c

(1) If at any time while a grant condition remains in force with respect to a dwelling, house or building—

(a) the owner of the dwelling, house or building to which the condition relates pays the amount of the grant to the local housing authority by whom the grant was made,

> (b) a mortgagee of the interest of the owner in that dwelling, house or building being a mortgagee entitled to exercise a power of sale, makes such a payment,
>
> (c) the local housing authority determine not to demand repayment on the breach of a grant condition, or
>
> (d) the authority demand repayment in whole or in part on the breach of a grant condition and that demand is satisfied,

that grant condition and any other grant conditions shall cease to be in force with respect to that dwelling, house or building.

(2) In the case of a grant condition imposed on a landlord's application for a common parts grant the references in subsection (1)(a) and (b) to the owner of the building are to the applicant or any such successor in title as is referred to in section 46(3).

(3) An amount paid by a mortgagee under subsection (1)(b) above shall be treated as part of the sums secured by the mortgage and may be discharged accordingly.

(4) The purposes authorised for the application of capital money by—

> (a) section 73 of the Settled Land Act 1925,
>
> (b) that section as applied by section 28 of the Law of Property Act 1925 in relation to trusts for sale, and
>
> (c) section 26 of the Universities and College Estates Act 1925,

include the making of payments under this section.

Supplementary provisions

56 Provisions relating to death of applicant

(1) References in this Chapter to the applicant, in relation to a grant or an application for a grant, shall be construed in relation to any time after his death as a reference to his personal representatives.

(2) Where the applicant dies after liability has been incurred for any preliminary or ancillary services or charges, the local housing authority may, if they think fit, pay grant in respect of some or all of those matters.

(3) Where the applicant dies after the relevant works have been begun and before the certified date, the local housing authority may, if they think fit, pay grant in respect of some or all of the works already carried out and other relevant works covered by the application.

(4) Nothing in this section shall be construed as preventing the provisions as to grant conditions applying in relation to any payment of grant under subsection (2) or (3).

57 Power of local housing authority to carry out works which would attract grant

(1) A local housing authority may by agreement with a person having the requisite interest execute at his expense—

> (a) any works towards the cost of which a grant under this Chapter is payable or might be paid on an application duly made and approved, and
>
> (b) any further works which it is in their opinion necessary or desirable to execute together with the works mentioned in paragraph (a).

(2) Except in the case of a common parts grant, the "requisite interest" means a qualifying owner's interest for the purposes of a renovation grant, or an owner's interest for the purposes of a disabled facilities grant or HMO grant, as the case may be.

(3) In the case of a common parts grant, the reference in subsection (1) to a person having the requisite interest is a reference to the person who has—

    (a) an owner's interest in the building, or

    (b) such an interest in a flat in the building as is mentioned in section 14(2)(a) to (d) (occupying tenants),

and has a power or duty to carry out the relevant works.

58 Minor definitions: Chapter I

In this Chapter—

"common parts", in relation to a building, includes the structure and exterior of the building and common facilities provided, whether in the building or elsewhere, for persons who include the occupiers of one or more flats in the building;

"common parts application", in relation to an application for a disabled facilities grant, means an application in respect of works to the common parts of a building containing one or more flats;

"conversion application"—

    (a) in relation to an application for a renovation grant, means an application in respect of works required for the provision of one or more dwellings by the conversion of a house or other building, and

    (b) in relation to an application for an HMO grant, means an application for a grant in respect of works for the provision of a house in multiple occupation by the conversion of a house or other building;

"flat" means a dwelling which is a separate set of premises, whether or not on the same floor, divided horizontally from some other part of the building.

59 Index of defined expressions: Chapter I

In this Chapter the expressions listed below are defined by or otherwise fall to be construed in accordance with the provisions indicated—

certificate of future occupation (in relation to an application for an HMO grant)	section 26(2)
certificate of intended letting (in relation to an application for a renovation grant)	section 8(3)
certified date	section 44(3)(b)
common parts	section 58
common parts application (in relation to a disabled facilities grant)	section 58
common parts grant	section 1(3)
connected (with the owner of a dwelling)	section 98(2)

conversion application	section 58
disabled facilities grant	section 1(4)
disabled occupant	section 20
disabled person	section 100(1) to (3)
dwelling	section 101
elderly	section 101
eligible works	section 34(2)(a)
estimated expense	section 34(2)
exempt disposal	section 54
fit for human habitation	section 97(1)
flat	section 58
grant (without more)	section 1(6)
grant condition	section 44(1)
grant condition period	section 44(3)(a)
HMO grant	section 1(5)
house in multiple occupation	section 101
housing action trust	section 101
improvement	section 101
introductory tenant	section 101
landlord's application	
–in relation to a renovation grant	section 10(6)
–in relation to a common parts grant	section 15(1) and (2)
local authority	section 101
local housing authority	section 101
long tenancy	section 101
member of family	section 98(1)
new town corporation	section 101
occupying tenant (in relation to an application for a common parts grant)	section 14(2)
owner	sections 99 and 101
owner-occupation certificate (in relation to an application for a renovation grant)	section 8(2)
owner's application	
–in relation to a renovation grant	section 7(1) and (2)
–in relation to a disabled facilities grant	section 19(1) and (2)

owner's certificate (in relation to an application for a disabled facilities grant)	section 21(2)
owner's interest	section 101
participating landlord (in relation to a tenants' application for a common parts grant)	section 15(4)
partner	section 101
preliminary or ancillary services and charges	section 2(3)
prescribed	section 101
qualifying owner's interest	
–in relation to an application for a renovation grant	section 7(4)
–in relation to an application for a disabled facilities grant	section 19(4)
–in relation to an application for an HMO grant	section 25(3)
qualifying tenant	
–in relation to an application for a renovation grant	section 7(5)
–in relation to an application for a disabled facilities grant	section 19(4)
reasonable repair	section 96
registered social landlord	section 101
relevant disposal	section 53
relevant works (in relation to a grant application)	section 2(2)(a)
renewal area	section 101
renovation grant	section 1(2)
secure tenancy and secure tenant	section 101
social services authority	section 100(4)
statutory tenancy and statutory tenant	section 101
tenancy and tenant (generally)	section 101
tenant (and expressions relating to tenancies)	
–in the context of a tenant's application for a renovation grant	section 7(6)
–in the context of a certificate of intended letting	section 8(4)

Status: *This is the original version (as it was originally enacted).*

–in the context of an application for a common parts grant	section 14(2)
–in the context of an application for disabled facilities grant	section 19(5)
tenant's application	
–in relation to a renovation grant	section 7(1) and (2)
–in relation to a disabled facilities grant	section 19(1) and (2)
tenants' application (in relation to a common parts grant)	section 15(1) and (2)
tenant's certificate	
–for the purposes of an application for a renovation grant	section 9(2)
–for the purposes of an application for a disabled facilities grant	section 22(2)
urban development corporation	section 101

CHAPTER II

GROUP REPAIR SCHEMES

Introductory

60   Group repair schemes

(1) A local housing authority may prepare a scheme (a "group repair scheme") for the carrying out of works—

    (a)   to put in reasonable repair the exterior of the buildings to which the scheme relates, or

    (b)   to render the buildings to which the scheme relates structurally stable,

or for both those purposes.

(2) For the purposes of this Chapter "building" includes the whole or part of a terrace of houses or other units.

(3) The scheme must satisfy the requirements of sections 61 and 62 as to the buildings to which it relates and the works specified in it.

61   Qualifying buildings

(1) The buildings to which a group repair scheme relates must be qualifying buildings.

(2) A building is a qualifying building if at the time the scheme is prepared it satisfies the conditions prescribed for qualifying buildings in relation to a group repair scheme.

(3) A group repair scheme must relate to at least one qualifying building which at the time the scheme is prepared satisfies the conditions prescribed for a primary building in relation to a group repair scheme.

(4) Each of the other qualifying buildings to which a group repair scheme relates must satisfy the conditions prescribed for an additional building in relation to a group repair scheme.

62 Scheme works

(1) The works specified in a group repair scheme ("scheme works") must be works of the following descriptions.

(2) In the case of works to put in reasonable repair the exterior of the buildings to which the scheme relates, the works must be—

    (a) works to the exterior of the buildings to which the scheme relates, or

    (b) so far only as may be necessary to give satisfactory effect to such works, additional works to other parts of the buildings,

and must be such that on completion of the works the exterior of the buildings will be in reasonable repair.

(3) In the case of works to render the buildings to which the scheme relates structurally stable, the works must be—

    (a) works to the structure or to the foundations of the buildings to which the scheme relates, or

    (b) other works necessary to give satisfactory effect to such works,

and must be such that on completion of the works the buildings will be structurally stable.

(4) For the purposes of this Chapter the exterior of a building means—

    (a) any part of the building which is exposed to the elements of wind and rain or otherwise faces into the open air (including, in particular, roofs, chimneys, walls, doors, windows, rainwater goods and external pipework), and

    (b) the curtilage of the building, including any wall within the curtilage which is constructed as a retaining wall or otherwise to protect the structure of the building.

(5) In relation to works to the curtilage of a building the reference in subsection (2)(b) to additional works to other parts of the building includes additional works on land outside the curtilage.

(6) For the purposes of this Chapter the exterior of a building shall not be regarded as in reasonable repair unless it is substantially free from rising or penetrating damp.

63 Approval of scheme by Secretary of State

(1) If a group repair scheme prepared by a local housing authority is approved by the Secretary of State, the authority may, with the consent of the persons participating in the scheme, enter into agreements to secure the carrying out of the works specified in the scheme.

(2) The approval of the Secretary of State may be given either to a specific scheme or generally to schemes which fulfil such criteria as he may from time to time specify.

(3) Different criteria may be specified for different types of scheme and for different areas.

(4) The approval of a scheme may be made conditional upon compliance with requirements specified by the Secretary of State.

Participation in group repair scheme

64   **Persons eligible to participate in group repair scheme**

(1) A person is eligible to participate in a group repair scheme if at the date of the approval of the scheme—

 (a) he has an owner's interest in a dwelling or other premises comprised in a building to which the scheme relates, and

 (b) as respects the dwelling or other premises in which he has an owner's interest he either—

  (i) is able to give possession of any part of the building to which scheme works are proposed to be carried out, or

  (ii) has the consent of the occupier of that part to the carrying out of those works.

In the case of a scheme not submitted for specific approval, the date of approval shall be taken to be the date on which the authority decide that the scheme fulfils the criteria for general approval.

(2) A person eligible to participate in a group repair scheme may participate as an assisted participant—

 (a) if the owner's interest which he has is an interest in a dwelling and he gives an owner-occupation certificate or a certificate of intended letting, or

 (b) if the owner's interest which he has is an interest in a house in multiple occupation and he gives a certificate of future occupation.

This is subject to the exceptions specified in subsection (7) or by order under that subsection.

(3) An "owner-occupation certificate" certifies that the person concerned—

 (a) has an owner's interest in the dwelling, and

 (b) intends that throughout the protected period he, or a member of his family, will live in the dwelling, as his (or that member's) only or main residence.

(4) A "certificate of intended letting" certifies that the person concerned—

 (a) has an owner's interest in the dwelling, and

 (b) intends that throughout the protected period the dwelling will be let or available for letting as a residence and not for a holiday to someone other than a member of his family.

In paragraph (b) "letting" does not include a letting on a long tenancy.

(5) In subsection (4) references to letting include the grant of a licence to occupy premises.

References in this Chapter to tenants, and other expressions relating to tenancies, in the context of a certificate of intended letting, shall be construed accordingly.

(6) A "certificate of future occupation" certifies that the person concerned—

 (a) has an owner's interest in the house, and

Status: *This is the original version (as it was originally enacted).*

    (b)   intends that throughout the protected period the house or a part of it (specified in the certificate) will be residentially occupied, or available for residential occupation, under tenancies or licences by persons who are not connected with the owner for the time being of the house.

In paragraph (b) "residential occupation" does not include occupation for a holiday, and "tenancies" does not include a long tenancy.

(7) The following may not participate in a group repair scheme as an assisted participant—

    (a)   a local authority;

    (b)   a new town corporation;

    (c)   the Development Board for Rural Wales;

    (d)   a health authority, special health authority or NHS trust;

    (e)   a police authority established under section 3 of the Police Act 1964;

    (f)   a housing action trust;

    (g)   a registered social landlord;

    (h)   any other authority, body or other person excluded by order of the Secretary of State.

(8) An order under subsection (7)(h) may proceed wholly or in part by reference to the provisions relating to entitlement to housing benefit, or any other form of assistance, as they have effect from time to time.

(9) A person eligible to participate in a group repair scheme who is unable to participate as an assisted participant may participate as an unassisted participant.

65   Scheme consent and restriction on works

(1) The persons who are eligible to participate in a group repair scheme do so by signifying consent ("scheme consent"), in accordance with the terms of the scheme, to the proposals to carry out the works specified in the scheme.

(2) No scheme works shall be carried out to a part of a building which consists of premises in respect of which no person eligible to participate has signified scheme consent, except as mentioned below.

(3) The restriction in subsection (2) does not apply to works carried out to premises in respect of which there is no person (or no ascertainable person) eligible to participate in the scheme.

(4) The restriction in subsection (2) does not apply to works—

    (a)   which are carried out to premises in respect of which the person eligible to participate consents to their being carried out but has not signified scheme consent (and, accordingly, is not liable to contribute), and

    (b)   which it is necessary to carry out in order satisfactorily to carry out any works specified in the scheme to another part of the same building in respect of which a person eligible to participate has signified scheme consent.

66   Certificate of completion date

(1) When the works specified in a group repair scheme are completed, the local housing authority shall send to each assisted participant a certificate specifying the date on which the works were completed to the authority's satisfaction.

(2) In this Chapter that date is referred to as "the completion date".

67 Contributions by participants

(1) The participants in a group repair scheme are liable to contribute to the cost, as notified to them under the scheme, of scheme works relating to the premises in which they have an interest, at a rate determined in accordance with this section.

(2) The cost of the works shall be apportioned between the several buildings and premises in such way as may be agreed between the participants with owner's interests in them or, in default of agreement, equally.

(3) In the case of an unassisted participant, the rate of contribution is 100 per cent.

(4) In the case of an assisted participant whose owner's interest is in premises other than a dwelling or house in multiple occupation, the rate of contribution is—

    (a) 25 per cent. where the building is in a renewal area, and

    (b) 50 per cent. in any other case.

The Secretary of State may by order amend paragraph (a) or (b) so as to specify a different percentage.

(5) In the case of any other assisted participant, the rate of contribution is a percentage determined by the local housing authority not exceeding that which would apply under subsection (4).

(6) In making their determination the authority shall have regard to the way in which—

    (a) section 30 (means-testing in case of application by owner-occupier or tenant), or

    (b) section 31 (determination of amount of grant in case of landlord's application),

would apply if he were an applicant for a renovation grant or, as the case may require, an HMO grant.

(7) They shall also have regard to any guidance given by the Secretary of State for the purposes of this section.

Different guidance may be given for different cases, different descriptions of cases and different areas and, in particular, with respect to different local housing authorities or descriptions of authority (including a description framed by reference to authorities in a particular area).

Variation of group repair scheme

68 Variation of group repair scheme

(1) A group repair scheme may be varied at any time before the completion date.

The variation may relate to the participants in the scheme, the buildings to which the scheme relates, the scheme works or any other matter.

(2) A variation is not effective unless approved by the Secretary of State.

The provisions of section 63(2) to (4) (supplementary provisions as to approval of scheme) apply to approval of a variation.

(3) Where a scheme is varied to enable other persons to participate, section 64 (persons eligible to participate) applies in relation to new participants with the substitution for the reference to the date of approval of the scheme of a reference to the date of approval of the variation.

In the case of a variation not submitted for specific approval, the date of approval shall be taken to be the date on which the authority decide that the variation fulfils the criteria for general approval.

(4) Before varying a group repair scheme the local housing authority shall consult the existing participants and consider any representations made by them.

(5) Fresh scheme consent is required in the case of an existing participant as to whom the authority are satisfied that his interests are adversely affected by the variation.

In any other case the existing scheme consent shall be treated as extended to the scheme as varied.

Conditions of participation

69   Conditions of participation: general

(1) The following sections have effect with respect to the conditions of participation in a group repair scheme as an assisted participant.

(2) Except as otherwise provided those conditions have effect for the period of five years, or such other period as may be prescribed, beginning with the completion date.

That period is referred to in this Chapter as "the protected period".

(3) For the purposes of those conditions the "balance of the cost" is the difference between—
  (a) the cost as notified to the participant under the scheme of such of the works specified in the scheme as relate to the premises in which his owner's interest subsisted, and
  (b) the amount of the contribution in respect of that cost paid by him by virtue of section 67.

70   Condition as to payment of balance of cost on disposal

(1) It is a condition of participation in a group repair scheme as an assisted participant that if, at any time after signifying scheme consent and before the end of the protected period, he makes a relevant disposal (other than an exempt disposal) of the premises in which he had an owner's interest at the date of the approval of the scheme, he shall pay to the local housing authority on demand the balance of the cost.

(2) The condition under this section is a local land charge and is binding on any person who is for the time being an owner of the premises concerned.

(3) Where the authority have the right to demand payment of an amount as mentioned in subsection (1), they may determine not to demand payment or to demand a lesser amount.

(4) The condition under this section shall cease to be in force with respect to any premises if there is a relevant disposal of the premises that is an exempt disposal, other than—

(a) a disposal within section 54(1)(a) (disposal to associates of person making disposal), or

(b) a disposal within section 54(1)(b) (vesting under will or on intestacy).

71 Conditions as to occupation

(1) It is a condition of participation in a group repair scheme as an assisted participant—

(a) where the participant gave an owner-occupation certificate, that throughout the protected period the dwelling is occupied in accordance with the intention stated in the certificate;

(b) where the participant gave a certificate of intended letting, that throughout the protected period the dwelling is let or available for letting in accordance with the intention stated in the certificate; and

(c) where the participant gave a certificate of future occupation, that throughout the protected period the house is residentially occupied, or available for residential occupation, in accordance with the intention stated in the certificate.

(2) It is also a condition of participation as an assisted participant that if at any time when any of the above conditions is in force the authority serve notice on the owner of the dwelling or house requiring him to do so, he will within the period of 21 days beginning with the date on which the notice was served furnish to the authority a statement showing how that condition is being fulfilled.

(3) A condition under this section is a local land charge and is binding on any person who is for the time being an owner of the dwelling or house.

(4) In the event of a breach of a condition under this section, the owner for the time being of the dwelling or house shall pay to the local housing authority on demand the balance of the cost.

(5) The local housing authority may determine not to make such a demand or may demand a lesser amount.

(6) Any condition under this section shall cease to be in force with respect to any premises if there is a relevant disposal of the premises which is an exempt disposal other than a disposal within section 54(1)(a) (disposal to associates of person making disposal).

72 Meaning of relevant disposal and exempt disposal

Sections 53 and 54 (meaning of "relevant disposal" and "exempt disposal") apply for the purposes of this Chapter.

73 Payment of balance of cost, &c: cessation of conditions

(1) If at any time while a condition of participation under section 70 or 71 remains in force—

(a) the assisted participant pays the balance of the cost to the local housing authority,

    (b)   a mortgagee of the interest of the assisted participant in the premises being a mortgagee entitled to exercise a power of sale, makes such a payment,

    (c)   the authority determine not to demand payment on the breach of a condition of participation, or

    (d)   the authority demand payment in whole or in part on the breach of a condition of participation and that demand is satisfied,

that condition and any other conditions of participation shall cease to be in force with respect to the premises of that assisted participant.

(2) An amount paid by a mortgagee under subsection (1)(b) above shall be treated as part of the sums secured by the mortgage and may be discharged accordingly.

(3) The purposes authorised for the application of capital money by—

    (a)   section 73 of the Settled Land Act 1925,

    (b)   that section as applied by section 28 of the Law of Property Act 1925 in relation to trusts for sale, and

    (c)   section 26 of the Universities and College Estates Act 1925,

include the making of payments under this section.

Supplementary provisions

74   **Power of Secretary of State to modify operation of Chapter**

(1) If the Secretary of State so directs in the case of any scheme or any description of scheme, such of the preceding provisions of this Chapter as are specified in the direction shall not apply in relation to that scheme or, as the case may be, in relation to a scheme of that description.

(2) The power under this section to give directions may be so exercised as to make different provision with respect to different local housing authorities or descriptions of authority (including a description framed by reference to authorities in a particular area).

75   **Index of defined expressions: Chapter II**

In this Chapter the expressions listed below are defined by or otherwise fall to be construed in accordance with the provisions indicated—

assisted participant	section 64(2) and (7)
balance of the cost (in relation to conditions of participation)	section 69(3)
building	section 60(2)
certificate of future occupation	section 64(6)
certificate of intended letting	section 64(4)
completion date	section 66(2)
connected (with the owner of a dwelling)	section 98(2)
dwelling	section 101

Status: *This is the original version (as it was originally enacted).*

eligible to participate	section 64(1)
exempt disposal	section 72 (and section 54)
group repair scheme	section 60(1)
house in multiple occupation	section 101
housing action trust	section 101
local authority	section 101
local housing authority	section 101
long tenancy	section 101
member of family	section 98(1)
new town corporation	section 101
owner	sections 99 and 101
owner-occupation certificate	section 64(3)
owner's interest	section 101
prescribed	section 101
protected period (in relation to conditions of participation)	section 69(2)
reasonable repair	section 96
registered social landlord	section 101
relevant disposal	section 72 (and section 53)
renewal area	section 101
scheme consent	section 65(1)
scheme works	section 62
tenancy and tenant (generally)	section 101
tenant and related expressions (in the context of a certificate of intended letting)	section 64(5)
unassisted participant	section 64(9)

CHAPTER III

HOME REPAIR ASSISTANCE

76   Home repair assistance

(1) A local housing authority may, on application being made to them, give assistance under this Chapter ("home repair assistance") in the form of a grant or the provision of materials for the carrying out of works of repair, improvement or adaptation to a dwelling.

(2) The Secretary of State may by order make provision as to the total amount or value of home repair assistance that may be given—

    (a) on any one application, or

    (b) in respect of the same dwelling in any period of three years.

(3) Home repair assistance shall not be given in respect of works—

    (a) for which a grant under Chapter I has been approved or in respect of which an application for a grant is pending, or

    (b) which are specified in a group repair scheme approved under Chapter II or prepared and awaiting the approval of the Secretary of State.

77   Entitlement to home repair assistance

(1) Subject to the following provisions of this section, a local housing authority shall not entertain an application for home repair assistance unless they are satisfied—

    (a) that the applicant is aged 18 or over on the date of the application,

    (b) that he lives in the dwelling as his only or main residence,

    (c) that he has an owner's interest in the dwelling, or is a tenant of the dwelling, alone or jointly with others,

    (d) that he has a duty or power to carry out the works in question, and

    (e) that he or his partner is in receipt of income support, family credit, housing benefit, council tax benefit or disability working allowance.

(2) In the case of an application in respect of works to adapt a dwelling to enable an elderly, disabled or infirm person to be cared for, the condition in subsection (1)(b) shall be treated as met if the elderly, disabled or infirm person (whether or not the applicant) lives or proposes to live in the dwelling as his only or main residence.

(3) For the purposes of the condition in subsection (1)(c) "tenant" includes—

    (a) a secure tenant or statutory tenant,

    (b) a protected occupier under the Rent (Agriculture) Act 1976 or a person in occupation under an assured agricultural occupancy within the meaning of Part I of the Housing Act 1988, and

    (c) an employee (whether full-time or part-time) who occupies the dwelling or flat concerned for the better performance of his duties;

but does not include a tenant of an authority or body mentioned in section 3(2) (authorities and bodies not eligible to apply for grants under Chapter I).

(4) An application may be made by a person who does not satisfy the condition in subsection (1)(c) but who occupies the dwelling under a right of exclusive occupation granted for his life or for a period of more than five years.

But except in the case of—

    (a) works to adapt a dwelling to enable an elderly, disabled or infirm person, who lives or proposes to live in the dwelling as his only or main residence, to be cared for,

    (b) works relating to means of escape from fire or other fire precautions, or

    (c) any works to a dwelling in a renewal area,

the local housing authority shall not entertain an application made by virtue of this subsection unless they are satisfied that the applicant has occupied the dwelling as his only or main residence for a period of at least three years immediately preceding the date of the application.

(5) The condition in subsection (1)(e) does not apply—

    (a)   to an applicant who is elderly, disabled or infirm, or

    (b)   to an application in respect of works to adapt a dwelling to enable an elderly, disabled or infirm person, who lives or proposes to live in the dwelling, to be cared for.

78   Assistance in respect of house-boats and mobile homes

(1) Subject to the following provisions of this section, sections 76 and 77 (home repair assistance) apply in relation to a house-boat or mobile home as in relation to a dwelling.

(2) For the purposes of those sections as they apply in relation to a house-boat or mobile home, any person lawfully in occupation of the house-boat or mobile home shall be treated as a person with an owner's interest in or a tenant of a dwelling.

But except in the case of—

    (a)   works to adapt a house-boat or mobile home to enable an elderly, disabled or infirm person, who lives or proposes to live there as his only or main residence, to be cared for, or

    (b)   works relating to means of escape from fire or other fire precautions,

the local housing authority shall not entertain an application for home repair assistance unless the residence requirement is met.

(3) The residence requirement in the case of a house-boat is that the local housing authority are satisfied that—

    (a)   the applicant has occupied the boat as his only or main residence for a period of at least three years immediately preceding the date of the application;

    (b)   the boat has for that period had its only or main mooring in the same locality on an inland waterway or in marine waters within the boundary of the authority; and

    (c)   the applicant had a right to moor his boat there.

(4) The residence requirement in the case of a mobile home is that the local housing authority are satisfied that—

    (a)   the applicant has occupied the mobile home as his only or main residence for a period of at least three years immediately preceding the date of the application;

    (b)   the mobile home has for that period been on land forming part of the same protected site within the meaning of the Mobile Homes Act 1983; and

    (c)   the applicant occupied it under an agreement to which that Act applies or under a gratuitous licence.

(5) In this section—

        "house-boat" means a boat or similar structure designed or adapted for use as a place of permanent habitation, and

        "mobile home" means a caravan within the meaning of Part I of the Caravan Sites and Control of Development Act 1960 (disregarding the amendment made by section 13(2) of the Caravan Sites Act 1968),

which is a dwelling for the purposes of Part I of the Local Government Finance Act 1992 (council tax).

79   Power to make further provision by regulations

(1) The Secretary of State may by regulations make provision as to—

    (a)   the manner of making an application for home repair assistance and the contents of such an application;

    (b)   the procedure for dealing with applications for home repair assistance and for ensuring that works are carried out to any standard specified in the regulations;

    (c)   the way in which the amount of home repair assistance to be given on any application is to be determined; and

    (d)   the taking into account (in such manner and to such extent as may be prescribed) of the financial circumstances of the applicant.

(2) The Secretary of State may by regulations make provision extending or restricting the availability of home repair assistance, by reference to such description of persons, circumstances or other factors as the Secretary of State thinks fit.

(3) Regulations under subsection (2) may proceed wholly or in part by reference to the provisions relating to entitlement to housing benefit, or any other form of assistance, as they have effect from time to time.

80   Index of defined expressions: Chapter III

In this Chapter the expressions listed below are defined by or otherwise fall to be construed in accordance with the provisions indicated—

disabled person	section 100(1) to (3)
dwelling	section 101
elderly	section 101
home repair assistance	section 76(1)
improvement	section 101
local housing authority	section 101
owner's interest	section 101
partner	section 101
prescribed	section 101
renewal area	section 101
secure tenant	section 101
statutory tenant	section 101
tenancy and tenant (generally)	section 101

CHAPTER IV

DEFERRED ACTION NOTICES, &C.

Deferred action notices

81   Deferred action notices

(1) If the local housing authority are satisfied that a dwelling-house or house in multiple occupation is unfit for human habitation, but are satisfied that serving a deferred action notice is the most satisfactory course of action, they shall serve such a notice.

(2) A deferred action notice is a notice—

- (a)   stating that the premises are unfit for human habitation,
- (b)   specifying the works which, in the opinion of the authority, are required to make the premises fit for human habitation, and
- (c)   stating the other courses of action which are available to the authority if the premises remain unfit for human habitation.

(3) The notice becomes operative, if no appeal is brought, on the expiry of 21 days from the date of the service of the notice and is final and conclusive as to matters which could have been raised on an appeal.

(4) A deferred action notice which has become operative is a local land charge so long as it remains operative.

(5) The fact that a deferred action notice has been served does not prevent the local housing authority from taking any other course of action in relation to the premises at any time.

82   Service of deferred action notices

(1) The local housing authority shall serve a deferred action notice—

- (a)   in the case of a notice relating to a dwelling-house, on the person having control of the dwelling-house as defined in section 207 of the Housing Act 1985;
- (b)   in the case of a notice relating to a house in multiple occupation, on the person having control of the house as defined in section 398 of that Act.

(2) Where the authority are satisfied that a dwelling-house which is a flat, or a flat in multiple occupation, is unfit for human habitation by virtue of section 604(2) of the Housing Act 1985, they shall also serve the notice on the person having control (as defined in section 207 of that Act) of the building or part of the building in question.

(3) In the case of a house in multiple occupation, the authority may serve the notice on the person managing the house instead of the person having control of the house.

(4) Where the authority serve a notice under subsection (1), (2) or (3)—

- (a)   they shall also serve a copy of the notice on any other person having an interest in the premises concerned, whether as freeholder, mortgagee or lessee (within the meaning of Part VI of the Housing Act 1985), and
- (b)   they may serve a copy of the notice on any person having a licence to occupy the premises.

(5) Section 617 of the Housing Act 1985 (service of notices) applies for the purpose of this section as it applies for the purpose of that Act.

83   Appeals against deferred action notices

(1) A person aggrieved by a deferred action notice may within 21 days after the service of the notice appeal to the county court.

(2) Without prejudice to the generality of subsection (1), it is a ground of appeal that serving a notice under section 189 of the Housing Act 1985, or making a closing order under section 264 of that Act or a demolition order under section 265 of that Act, is a more satisfactory course of action.

(3) Where the grounds on which an appeal is brought are or include that specified in subsection (2), the court, on the hearing of the appeal, shall have regard to any guidance given to the local housing authority under section 604A of the Housing Act 1985 or section 85 of this Act.

(4) On an appeal the court may make such order either confirming, quashing or varying the notice as it thinks fit.

(5) Where the appeal is allowed and the reason or one of the reasons for allowing the appeal is that serving a notice under section 189 of that Act or making a closing order under section 264 of that Act or a demolition order under section 265 of that Act is a more satisfactory course of action, the judge shall, if requested to do so by the appellant or the local housing authority, include in his judgment a finding to that effect.

(6) If an appeal is brought, the deferred action notice does not become operative until—

   (a)   a decision on the appeal confirming the notice (with or without variation) is given and the period within which an appeal to the Court of Appeal may be brought expires without any such appeal having been brought, or

   (b)   if a further appeal to the Court of Appeal is brought, a decision on that appeal is given confirming the notice (with or without variation);

and for this purpose the withdrawal of an appeal has the same effect as a decision confirming the notice or decision appealed against.

84   Review of deferred action notices

(1) The local housing authority may at any time review any deferred action notice served by them, and they shall do so not later than two years after the notice becomes operative and at intervals of not more than two years thereafter.

   The Secretary of State may by order amend this subsection so as to specify such other period or periods as he considers appropriate.

(2) The authority shall for the purposes of any such review inspect the premises concerned.

   For this purpose sections 197 (powers of entry) and 198 (penalty for obstruction) of the Housing Act 1985 apply as they apply for the purposes of Part VI of that Act.

(3) If the authority are satisfied that the deferred action notice remains the most satisfactory course of action, they shall renew the notice and serve notice of their decision.

(4) The provisions of section 82 (service of deferred action notice) and section 83(1) to (5) (appeals against deferred action notices) apply in relation to the authority's decision to renew a deferred action notice as in relation to the original notice.

(5) If an appeal is brought against the decision to renew a deferred action notice, the notice remains operative until any decision on the appeal, or any further appeal, quashing or varying the notice.

(6) If the authority take action in relation to the premises under any of the provisions listed in section 604A(1) of the Housing Act 1985, the deferred action notice shall cease to be operative on the relevant notice, order or declaration becoming operative.

85 Guidance by Secretary of State

(1) In deciding for the purposes of section 81 (deferred action notices) or section 84 (review of deferred action notices) what is the most satisfactory course of action in relation to any premises, the local housing authority shall have regard to such guidance as may from time to time be given by the Secretary of State.

(2) The provisions of section 604A(2) to (4) of the Housing Act 1985 (supplementary provisions as to guidance) apply in relation to such guidance.

Power to improve enforcement procedures

86 Unfitness for human habitation &c.: power to improve enforcement procedures

(1) The Secretary of State may by order provide that a local housing authority shall act as specified in the order before taking action of any of the following kinds—

  (a) serving a deferred action notice under section 81 or renewing such a notice under section 84;

  (b) serving a notice under section 189 of the Housing Act 1985 (repair notice in respect of house which unfit for human habitation);

  (c) serving a notice under section 190 of that Act (repair notice in respect of house in state of disrepair but not unfit for human habitation);

  (d) making a closing order under section 264 of that Act;

  (e) making a demolition order under section 265 of that Act.

(2) An order under this section may provide that the authority—

  (a) shall as soon as practicable give to the person against whom action is intended a written notice which satisfies the requirements of subsection (3); and

  (b) shall not take any action against him until after the end of such period beginning with the giving of the notice as may be determined by or under the order.

(3) A notice satisfies the requirements of this subsection if it—

  (a) states the nature of the remedial action which in the authority's opinion should be taken, and explains why and within what period;

  (b) explains the grounds on which it appears to the authority that action might be taken as mentioned in subsection (1); and

  (c) states the nature of the action which could be taken and states whether there is a right to make representations before, or a right of appeal against, the taking of such action.

(4) An order under this section may also provide that, before the authority takes any action against any person, they—

  (a) shall give to that person a written notice stating—

    (i) that they are considering taking the action and the reasons why they are considering it; and

    (ii) that the person may, within a period specified in the notice, make written representations to them or, if the person so requests, make oral representations to them in the presence of a person determined by or under the order; and

  (b) shall consider any representations which are duly made and not withdrawn.

(5) An order under this section may in particular—

  (a) make provision as to the consequences of any failure to comply with a provision made by the order;

  (b) contain provisions (including provisions modifying enactments relating to the periods within which proceedings must be brought) which are consequential upon, or supplemental or incidental to, the provisions made by the order.

(6) Nothing in any order made under this section shall—

  (a) preclude a local housing authority from taking immediate action against any person, or from requiring any person to take immediate remedial action to avoid action being taken against him, in any case where it appears to them to be necessary to take such action or impose such a requirement; or

  (b) require such an authority to disclose any information the disclosure of which would be contrary to the public interest.

Power to charge for enforcement action

87 Unfitness for human habitation, &c.: power to charge for enforcement action

(1) A local housing authority may make such reasonable charge as they consider appropriate as a means of recovering certain administrative and other expenses incurred by them in taking action of any of the following kinds—

  (a) serving a deferred action notice under section 81 or deciding to renew such a notice under section 84;

  (b) serving a notice under section 189 of the Housing Act 1985 (repair notice in respect of house which unfit for human habitation);

  (c) serving a notice under section 190 of that Act (repair notice in respect of house in state of disrepair but not unfit for human habitation);

  (d) making a closing order under section 264 of that Act;

  (e) making a demolition order under section 265 of that Act.

(2) The expenses are, in the case of the service of a notice under section 81 of this Act or section 189 or 190 of the Housing Act 1985, the expenses incurred in—

  (a) determining whether to serve the notice,

  (b) identifying the works to be specified in the notice, and

  (c) serving the notice.

(3) The expenses are, in the case of a decision to renew a notice under section 84 of this Act, the expenses incurred in—

    (a) deciding whether to renew the notice, and

    (b) serving notice of the authority's decision.

(4) The expenses are, in the case of a closing order under section 264 of the Housing Act 1985 or a demolition order under section 265 of that Act, the expenses incurred in—

    (a) determining whether to make the order, and

    (b) serving notice of the order.

(5) The amount of the charge shall not exceed such amount as is specified by order of the Secretary of State.

(6) Where a court allows an appeal against the underlying notice, decision or order mentioned in subsection (1), it may make such order as it thinks fit reducing, quashing or requiring the repayment of any charge under this section made in respect of the notice, decision or order.

88   Recovery of charge for enforcement action

(1) The following provisions have effect with respect to the recovery of a charge under section 87.

(2) The charge may be recovered by the authority concerned from—

    (a) in the case of a notice under section 81 of this Act, or section 189 or 190 of the Housing Act 1985, any person on whom the notice is served;

    (b) in the case of a renewal of a notice under section 84 of this Act, any person on whom notice of the decision to renew the notice is served;

    (c) in the case of an order under section 264 or 265 of the Housing Act 1985, any person on whom notice of the order is served as an owner of the premises.

(3) A demand for payment of the charge shall be served on the person from whom the authority seeks to recover it.

(4) The demand becomes operative, if no appeal is brought against the underlying notice, decision or order, on the expiry of the period of 21 days from the service of the demand.

(5) The sum recoverable by the authority is, until recovered, a charge on the premises concerned; and—

    (a) the charge takes effect when the demand becomes operative,

    (b) the authority have for the purpose of enforcing the charge the same powers and remedies under the Law of Property Act 1925 and otherwise as if they were mortgagees by deed having powers of sale and lease, of accepting surrenders of leases and of appointing a receiver, and

    (c) the power of appointing a receiver is exercisable at any time after the expiration of one month from the date when the charge takes effect.

Supplementary provisions

89 **Power to prescribe forms**

The Secretary of State may by regulations prescribe the form of and the particulars to be contained in—

(a) a deferred action notice, or a notice of an authority's decision to renew a deferred action notice, or

(b) a demand for payment of any charge under section 87 (power to charge for enforcement action).

90 **Minor definitions: Chapter IV**

In this Chapter—

(a) "dwelling-house", "flat" and references to the owner of a dwelling-house or flat, have the same meaning as in Part VI of the Housing Act 1985 (repair notices); and

(b) "house in multiple occupation", "flat in multiple occupation" and references to the owner of or person managing such a house or flat, have the same meaning as in Part XI of that Act.

91 **Index of defined expressions: Chapter IV**

In this Chapter the expressions listed below are defined by or otherwise fall to be construed in accordance with the provisions indicated—

deferred action notice	section 81
dwelling-house	section 90(a)
flat	section 90(a)
flat in multiple occupation	section 90(b)
house in multiple occupation	section 90(b)
local housing authority	section 101
owner	
–in relation to a dwelling-house or flat	section 90(a)
–in relation to a house or flat in multiple occupation	section 90(b)
person managing (a house or flat in multiple occupation)	section 90(b)
prescribed	section 101
unfit for human habitation	section 97

Status: *This is the original version (as it was originally enacted).*

CHAPTER V

SUPPLEMENTARY PROVISIONS

Contributions by Secretary of State

92   Contributions by the Secretary of State

(1) The Secretary of State may pay contributions to local housing authorities towards such expenditure incurred by them under this Part as he may determine.

(2) The rate or rates of the contributions, the calculation of the expenditure to which they relate and the manner of their payment shall be such as may be determined by the Secretary of State with the consent of the Treasury.

(3) A determination under subsection (1) or (2)—

    (a)   may be made generally or with respect to a particular local housing authority or description of authority, including a description framed by reference to authorities in a particular area, and

    (b)   may make different provision in relation to different cases or descriptions of case.

(4) Contributions under this section shall be payable subject to such conditions as to records, certificates, audit or otherwise as the Secretary of State may, with the approval of the Treasury, impose.

(5) If, before the declaration of a renewal area, a local housing authority are satisfied that the rate of contributions which, in accordance with a determination under subsection (2), would otherwise be applicable to the authority will not be adequate, bearing in mind the action they propose to take with regard to the area, they may, before making the declaration, apply to the Secretary of State for contributions at a higher rate in respect of that area.

(6) An application under subsection (5) shall be made in such form and shall contain such particulars as the Secretary of State may determine; and, if such an application is made, the authority shall not declare the area concerned to be a renewal area until the application is approved, refused or withdrawn.

(7) If an application under subsection (5) is approved, the Secretary of State may pay contributions under subsection (1) in respect of the area concerned at such higher rate as he may determine under subsection (2).

93   Recovery of contributions

(1) Where the Secretary of State has paid contributions under section 92 to a local housing authority, he may recover from the authority such amount as he determines to be appropriate in respect of repayments of grant under this Part.

(2) For the purposes of this section—

    (a)   a "grant" includes the cost of scheme works for a group repair scheme (see section 62(1)), and

    (b)   "repayment of grant" includes the payment to the authority of the balance of the cost (see section 69(3)) by assisted participants in such a scheme.

(3) The amount shall be calculated by reference to the amount appearing to the Secretary of State to represent his contribution to—

    (a)   grants in respect of which repayments have been made to the authority, or

    (b)   grants in respect of which repayments could have been recovered if reasonable steps had been taken by the authority,

together with an appropriate percentage of any interest received by the authority, or which would have been received if reasonable steps had been taken by the authority.

(4) The question what steps it would have been reasonable for the authority to take shall be determined by the Secretary of State.

In determining whether the authority took reasonable steps, the Secretary of State may consider whether the authority properly exercised its discretion not to demand repayment of grant or to demand payment of a lesser sum.

Consent of the Secretary of State

94   Consent of the Secretary of State

The consent of the Secretary of State for the purposes of—

    (a)   section 45(4)(b), 46(4) or 47(4) (consent to waiver of liability to repay renovation grant, common parts grant or HMO grant on disposal), or

    (b)   section 34(6)(b), 44(3)(a), 51 or 52 (conditions imposed with consent of Secretary of State),

may be given either generally or in relation to any one or more specified authorities or descriptions of authority or in relation to particular cases or descriptions of case.

Parsonages, charities, &c.

95   Parsonages, charities, &c

(1) The provisions of Chapter I (main grants) mentioned below do not apply to—

    (a)   an application for a grant in respect of glebe land or the residence house of an ecclesiastical benefice, or

    (b)   an application for a grant made by a charity or on behalf of a charity by the charity trustees of the charity.

(2) Those provisions are—

    (a)   sections 7 to 11 (conditions for application for renovation grant);

    (b)   sections 19, 21 and 22 (conditions for application for disabled facilities grant);

    (c)   sections 25 and 26 (conditions for application for HMO grant).

(3) In considering under section 31 the amount (if any) of the grant where the applicant is a charity or the application is in respect of glebe land, the local housing authority shall have regard, in addition to the matters mentioned in that section, to any obligation or practice on the part of the applicant to let dwellings at a rent less than that which could be obtained on the open market.

(4) In Chapter II (group repair schemes), in section 64(2) (persons eligible to participate in group repair scheme as assisted participants), the requirement in paragraph (a) that

a person give an owner-occupation certificate or a certificate of intended letting does not apply if—

    (a)   the person concerned is a charity or the trustee of a charity, or

    (b)   the dwelling is the residence house of an ecclesiastical benefice;

and the requirement in paragraph (b) that a person give a certificate of future occupation does not apply if the person concerned is a charity or the trustee of a charity.

(5) In Chapter III (home repair assistance), section 77(1)(c) (condition that applicant have owner's interest or tenancy) does not apply to an application by an individual in respect of glebe land or the residence house of an ecclesiastical benefice.

(6) In this section "charity" does not include a registered social landlord but otherwise has the same meaning as in the Charities Act 1993.

Interpretation

96   Meaning of "reasonable repair"

In determining for the purposes of this Part what is "reasonable repair", in relation to a dwelling, house or building, a local housing authority—

    (a)   shall have regard to the age and character of the dwelling, house or building and the locality in which it is situated, and

    (b)   shall disregard the state of internal decorative repair.

97   Fitness for human habitation

(1) Section 604 of the Housing Act 1985 (fitness for human habitation) applies for the purposes of this Part as it applies for the purposes of that Act.

(2) In deciding whether they are satisfied that the carrying out of the relevant works is the most satisfactory course of action in a case where the house or dwelling concerned is unfit for human habitation, the local housing authority shall have regard to any guidance given under section 604A of the Housing Act 1985 and section 85 of this Act.

For that purpose the authority shall treat any guidance given in respect of the serving of a repair notice under section 189(1) of the Housing Act 1985 as guidance given in respect of the completion of the relevant works.

98   Members of a person's family and connected persons

(1) Section 113 of the Housing Act 1985 (meaning of "members of a person's family") applies in determining whether a person is a member of another's family for the purposes of this Part.

(2) For the purposes of this Part a person is connected with the owner for the time being of a dwelling if—

    (a)   in a case where personal representatives or trustees are the owner, he is a person who under the will or intestacy or, as the case may be, under the terms of the trust concerned is beneficially entitled to an interest in the dwelling or to the proceeds of sale of the dwelling;

    (b)   in any other case, he is a member of the family of the owner.

99   Meaning of "owner" of dwelling

(1) In this Part "owner", in relation to a dwelling, means the person who—

    (a)   is for the time being entitled to receive from a lessee of the dwelling (or would be so entitled if the dwelling were let) a rent at an annual rate of not less than two-thirds of the net annual value of the dwelling; and

    (b)   is not himself liable as lessee of the dwelling, or of property which includes the dwelling, to pay such a rent to a superior landlord.

(2) For this purpose the net annual value of a dwelling means the rent at which the dwelling might reasonably be expected to be let from year to year if the tenant undertook to pay all usual tenant's rates and taxes and to bear the cost of repair and insurance and the other expenses, if any, necessary to maintain the dwelling in a state to command that rent.

(3) Any dispute arising as to the net annual value of a dwelling shall be referred in writing for decision by the district valuer.

In this subsection "district valuer" has the same meaning as in the Housing Act 1985.

100   Disabled persons

(1) For the purposes of this Part a person is disabled if—

    (a)   his sight, hearing or speech is substantially impaired,

    (b)   he has a mental disorder or impairment of any kind, or

    (c)   he is physically substantially disabled by illness, injury, impairment present since birth, or otherwise.

(2) A person aged eighteen or over shall be taken for the purposes of this Part to be disabled if—

    (a)   he is registered in pursuance of any arrangements made under section 29(1) of the National Assistance Act 1948 (disabled persons' welfare), or

    (b)   he is a person for whose welfare arrangements have been made under that provision or, in the opinion of the social services authority, might be made under it.

(3) A person under the age of eighteen shall be taken for the purposes of this Part to be disabled if—

    (a)   he is registered in a register of disabled children maintained under paragraph 2 of Schedule 2 to the Children Act 1989, or

    (b)   he is in the opinion of the social services authority a disabled child as defined for the purposes of Part III of the Children Act 1989 (local authority support for children and their families).

(4) In this Part the "social services authority" means the council which is the local authority for the purposes of the Local Authority Social Services Act 1970 for the area in which the dwelling or building is situated.

(5) Nothing in subsection (1) above shall be construed as affecting the persons who are to be regarded as disabled under section 29(1) of the National Assistance Act 1948 or section 17(11) of the Children Act 1989 (which define disabled persons for the purposes of the statutory provisions mentioned in subsections (2) to (4) above).

101   Minor definitions: Part I

In this Part—

"dwelling" means a building or part of a building occupied or intended to be occupied as a separate dwelling, together with any yard, garden, outhouses and appurtenances belonging to it or usually enjoyed with it;

"elderly" means aged 60 years or over;

"house in multiple occupation" has the same meaning as in Part VII of the Local Government and Housing Act 1989;

"housing action trust" means a housing action trust established under Part III of the Housing Act 1988 and includes any body established by order under section 88 of the Housing Act 1988;

"improvement" includes alteration and enlargement;

"introductory tenancy" and "introductory tenant" have the same meaning as in Chapter I of Part V of the Housing Act 1996;

"local authority" and "local housing authority" have the same meaning as in the Housing Act 1985;

"long tenancy" has the meaning assigned by section 115 of that Act;

"new town corporation" has the same meaning as in the Housing Act 1985 and includes any body established by order under paragraph 7 of Schedule 9 to the New Towns Act 1981;

"owner", in relation to a dwelling, has the meaning given by section 99, and, in relation to a house in multiple occupation, has the same meaning as in Part XI of the Housing Act 1985;

"owner's interest", in relation to any premises, means—

(a) an estate in fee simple absolute in possession, or

(b) a term of years absolute of which not less than five years remain unexpired at the date of the application,

whether held by the applicant alone or jointly with others;

"partner", in relation to a person, means that person's spouse or a person other than a spouse with whom he or she lives as husband or wife;

"prescribed" means prescribed by regulations made by the Secretary of State;

"registered social landlord" has the same meaning as in Part I of the Housing Act 1996;

"renewal area" has the same meaning as in Part VII of the Local Government and Housing Act 1989;

"secure tenancy" and "secure tenant" have the same meaning as in Part IV of the Housing Act 1985;

"statutory tenancy" and "statutory tenant" mean a statutory tenancy or statutory tenant within the meaning of the Rent Act 1977 or the Rent (Agriculture) Act 1976;

"tenancy" includes a sub-tenancy and an agreement for a tenancy or sub-tenancy;

"tenant" includes a sub-tenant and any person deriving title under the original tenant or sub-tenant;

"urban development corporation" has the same meaning as in the Housing Act 1985 and includes any body established by order under section 165B of the Local Government, Planning and Land Act 1980.

Transitional and consequential provisions

102   Transitional provisions

(1) The provisions of Chapters I to III of this Part have effect in place of Part VIII of the Local Government and Housing Act 1989 (grants towards cost of improvements and repairs, &c.).

(2) Subject as follows, the provisions of that Part continue to apply to applications for grant of the descriptions mentioned in section 101 of that Act made before the commencement of this Part.

(3) Sections 112 and 113 of that Act (which require a local housing authority to approve certain grant applications) do not apply to an application under that Part made after 2nd February 1996 which has not been approved or refused before the commencement of this Part, unless—

   (a)   the six month period under section 116(1) of that Act (period within which applicant to be notified of decision) has elapsed before commencement, or

   (b)   the works were begun on or before 2nd February 1996—

         (i) in an emergency, or

         (ii) in order to comply with a notice under section 189, 190 or 352 of the Housing Act 1985.

(4) An application to which section 112 or 113 of the Local Government and Housing Act 1989 would have applied but for subsection (3) above shall be dealt with after the commencement of this Part as if those sections were omitted from Part VIII of that Act.

(5) The above provisions do not affect the power conferred by section 150(4) to make transitional provision and savings in relation to the commencement of this Part, including provision supplementary or incidental to the above provisions.

Supplementary and incidental provision may, in particular, be made adapting the provisions of Part VIII of that Act in the case of applications to which section 112 or 113 would have applied but for the above provisions.

103   Consequential amendments: Part I

The enactments mentioned in Schedule 1 have effect with the amendments specified there which are consequential on the provisions of this Part.

PART II

CONSTRUCTION CONTRACTS

Introductory provisions

104   Construction contracts

(1) In this Part a "construction contract" means an agreement with a person for any of the following—

   (a)   the carrying out of construction operations;

    (b)   arranging for the carrying out of construction operations by others, whether under sub-contract to him or otherwise;

    (c)   providing his own labour, or the labour of others, for the carrying out of construction operations.

(2) References in this Part to a construction contract include an agreement—

    (a)   to do architectural, design, or surveying work, or

    (b)   to provide advice on building, engineering, interior or exterior decoration or on the laying-out of landscape,

in relation to construction operations.

(3) References in this Part to a construction contract do not include a contract of employment (within the meaning of the Employment Rights Act 1996).

(4) The Secretary of State may by order add to, amend or repeal any of the provisions of subsection (1), (2) or (3) as to the agreements which are construction contracts for the purposes of this Part or are to be taken or not to be taken as included in references to such contracts.

No such order shall be made unless a draft of it has been laid before and approved by a resolution of each of House of Parliament.

(5) Where an agreement relates to construction operations and other matters, this Part applies to it only so far as it relates to construction operations.

An agreement relates to construction operations so far as it makes provision of any kind within subsection (1) or (2).

(6) This Part applies only to construction contracts which—

    (a)   are entered into after the commencement of this Part, and

    (b)   relate to the carrying out of construction operations in England, Wales or Scotland.

(7) This Part applies whether or not the law of England and Wales or Scotland is otherwise the applicable law in relation to the contract.

105   Meaning of "construction operations"

(1) In this Part "construction operations" means, subject as follows, operations of any of the following descriptions—

    (a)   construction, alteration, repair, maintenance, extension, demolition or dismantling of buildings, or structures forming, or to form, part of the land (whether permanent or not);

    (b)   construction, alteration, repair, maintenance, extension, demolition or dismantling of any works forming, or to form, part of the land, including (without prejudice to the foregoing) walls, roadworks, power-lines, telecommunication apparatus, aircraft runways, docks and harbours, railways, inland waterways, pipe-lines, reservoirs, water-mains, wells, sewers, industrial plant and installations for purposes of land drainage, coast protection or defence;

    (c)   installation in any building or structure of fittings forming part of the land, including (without prejudice to the foregoing) systems of heating, lighting, air-conditioning, ventilation, power supply, drainage, sanitation, water supply or fire protection, or security or communications systems;

> (d) external or internal cleaning of buildings and structures, so far as carried out in the course of their construction, alteration, repair, extension or restoration;
>
> (e) operations which form an integral part of, or are preparatory to, or are for rendering complete, such operations as are previously described in this subsection, including site clearance, earth-moving, excavation, tunnelling and boring, laying of foundations, erection, maintenance or dismantling of scaffolding, site restoration, landscaping and the provision of roadways and other access works;
>
> (f) painting or decorating the internal or external surfaces of any building or structure.

(2) The following operations are not construction operations within the meaning of this Part—

> (a) drilling for, or extraction of, oil or natural gas;
>
> (b) extraction (whether by underground or surface working) of minerals; tunnelling or boring, or construction of underground works, for this purpose;
>
> (c) assembly, installation or demolition of plant or machinery, or erection or demolition of steelwork for the purposes of supporting or providing access to plant or machinery, on a site where the primary activity is—
>
> > (i) nuclear processing, power generation, or water or effluent treatment, or
> >
> > (ii) the production, transmission, processing or bulk storage (other than warehousing) of chemicals, pharmaceuticals, oil, gas, steel or food and drink;
>
> (d) manufacture or delivery to site of—
>
> > (i) building or engineering components or equipment,
> >
> > (ii) materials, plant or machinery, or
> >
> > (iii) components for systems of heating, lighting, air-conditioning, ventilation, power supply, drainage, sanitation, water supply or fire protection, or for security or communications systems,
>
> except under a contract which also provides for their installation;
>
> (e) the making, installation and repair of artistic works, being sculptures, murals and other works which are wholly artistic in nature.

(3) The Secretary of State may by order add to, amend or repeal any of the provisions of subsection (1) or (2) as to the operations and work to be treated as construction operations for the purposes of this Part.

(4) No such order shall be made unless a draft of it has been laid before and approved by a resolution of each House of Parliament.

106 Provisions not applicable to contract with residential occupier

(1) This Part does not apply—

> (a) to a construction contract with a residential occupier (see below), or
>
> (b) to any other description of construction contract excluded from the operation of this Part by order of the Secretary of State.

(2) A construction contract with a residential occupier means a construction contract which principally relates to operations on a dwelling which one of the parties to the contract occupies, or intends to occupy, as his residence.

In this subsection "dwelling" means a dwelling-house or a flat; and for this purpose—

"dwelling-house" does not include a building containing a flat; and

"flat" means separate and self-contained premises constructed or adapted for use for residential purposes and forming part of a building from some other part of which the premises are divided horizontally.

(3) The Secretary of State may by order amend subsection (2).

(4) No order under this section shall be made unless a draft of it has been laid before and approved by a resolution of each House of Parliament.

107   Provisions applicable only to agreements in writing

(1) The provisions of this Part apply only where the construction contract is in writing, and any other agreement between the parties as to any matter is effective for the purposes of this Part only if in writing.

The expressions "agreement", "agree" and "agreed" shall be construed accordingly.

(2) There is an agreement in writing—
  (a) if the agreement is made in writing (whether or not it is signed by the parties),
  (b) if the agreement is made by exchange of communications in writing, or
  (c) if the agreement is evidenced in writing.

(3) Where parties agree otherwise than in writing by reference to terms which are in writing, they make an agreement in writing.

(4) An agreement is evidenced in writing if an agreement made otherwise than in writing is recorded by one of the parties, or by a third party, with the authority of the parties to the agreement.

(5) An exchange of written submissions in adjudication proceedings, or in arbitral or legal proceedings in which the existence of an agreement otherwise than in writing is alleged by one party against another party and not denied by the other party in his response constitutes as between those parties an agreement in writing to the effect alleged.

(6) References in this Part to anything being written or in writing include its being recorded by any means.

Adjudication

108   Right to refer disputes to adjudication

(1) A party to a construction contract has the right to refer a dispute arising under the contract for adjudication under a procedure complying with this section.

For this purpose "dispute" includes any difference.

(2) The contract shall—
  (a) enable a party to give notice at any time of his intention to refer a dispute to adjudication;
  (b) provide a timetable with the object of securing the appointment of the adjudicator and referral of the dispute to him within 7 days of such notice;

> (c) require the adjudicator to reach a decision within 28 days of referral or such longer period as is agreed by the parties after the dispute has been referred;
>
> (d) allow the adjudicator to extend the period of 28 days by up to 14 days, with the consent of the party by whom the dispute was referred;
>
> (e) impose a duty on the adjudicator to act impartially; and
>
> (f) enable the adjudicator to take the initiative in ascertaining the facts and the law.

(3) The contract shall provide that the decision of the adjudicator is binding until the dispute is finally determined by legal proceedings, by arbitration (if the contract provides for arbitration or the parties otherwise agree to arbitration) or by agreement.

The parties may agree to accept the decision of the adjudicator as finally determining the dispute.

(4) The contract shall also provide that the adjudicator is not liable for anything done or omitted in the discharge or purported discharge of his functions as adjudicator unless the act or omission is in bad faith, and that any employee or agent of the adjudicator is similarly protected from liability.

(5) If the contract does not comply with the requirements of subsections (1) to (4), the adjudication provisions of the Scheme for Construction Contracts apply.

(6) For England and Wales, the Scheme may apply the provisions of the Arbitration Act 1996 with such adaptations and modifications as appear to the Minister making the scheme to be appropriate.

For Scotland, the Scheme may include provision conferring powers on courts in relation to adjudication and provision relating to the enforcement of the adjudicator's decision.

Payment

109   Entitlement to stage payments

(1) A party to a construction contract is entitled to payment by instalments, stage payments or other periodic payments for any work under the contract unless—

> (a) it is specified in the contract that the duration of the work is to be less than 45 days, or
>
> (b) it is agreed between the parties that the duration of the work is estimated to be less than 45 days.

(2) The parties are free to agree the amounts of the payments and the intervals at which, or circumstances in which, they become due.

(3) In the absence of such agreement, the relevant provisions of the Scheme for Construction Contracts apply.

(4) References in the following sections to a payment under the contract include a payment by virtue of this section.

110   Dates for payment

(1) Every construction contract shall—

    (a)   provide an adequate mechanism for determining what payments become due under the contract, and when, and

    (b)   provide for a final date for payment in relation to any sum which becomes due.

The parties are free to agree how long the period is to be between the date on which a sum becomes due and the final date for payment.

(2) Every construction contract shall provide for the giving of notice by a party not later than five days after the date on which a payment becomes due from him under the contract, or would have become due if—

    (a)   the other party had carried out his obligations under the contract, and

    (b)   no set-off or abatement was permitted by reference to any sum claimed to be due under one or more other contracts,

specifying the amount (if any) of the payment made or proposed to be made, and the basis on which that amount was calculated.

(3) If or to the extent that a contract does not contain such provision as is mentioned in subsection (1) or (2), the relevant provisions of the Scheme for Construction Contracts apply.

111   Notice of intention to withhold payment

(1) A party to a construction contract may not withhold payment after the final date for payment of a sum due under the contract unless he has given an effective notice of intention to withhold payment.

The notice mentioned in section 110(2) may suffice as a notice of intention to withhold payment if it complies with the requirements of this section.

(2) To be effective such a notice must specify—

    (a)   the amount proposed to be withheld and the ground for withholding payment, or

    (b)   if there is more than one ground, each ground and the amount attributable to it,

and must be given not later than the prescribed period before the final date for payment.

(3) The parties are free to agree what that prescribed period is to be.

In the absence of such agreement, the period shall be that provided by the Scheme for Construction Contracts.

(4) Where an effective notice of intention to withhold payment is given, but on the matter being referred to adjudication it is decided that the whole or part of the amount should be paid, the decision shall be construed as requiring payment not later than—

    (a)   seven days from the date of the decision, or

    (b)   the date which apart from the notice would have been the final date for payment,

whichever is the later.

112   Right to suspend performance for non-payment

(1) Where a sum due under a construction contract is not paid in full by the final date for payment and no effective notice to withhold payment has been given, the person to whom the sum is due has the right (without prejudice to any other right or remedy)

to suspend performance of his obligations under the contract to the party by whom payment ought to have been made ("the party in default").

(2) The right may not be exercised without first giving to the party in default at least seven days' notice of intention to suspend performance, stating the ground or grounds on which it is intended to suspend performance.

(3) The right to suspend performance ceases when the party in default makes payment in full of the amount due.

(4) Any period during which performance is suspended in pursuance of the right conferred by this section shall be disregarded in computing for the purposes of any contractual time limit the time taken, by the party exercising the right or by a third party, to complete any work directly or indirectly affected by the exercise of the right.

Where the contractual time limit is set by reference to a date rather than a period, the date shall be adjusted accordingly.

113   Prohibition of conditional payment provisions

(1) A provision making payment under a construction contract conditional on the payer receiving payment from a third person is ineffective, unless that third person, or any other person payment by whom is under the contract (directly or indirectly) a condition of payment by that third person, is insolvent.

(2) For the purposes of this section a company becomes insolvent—
   (a)   on the making of an administration order against it under Part II of the Insolvency Act 1986,
   (b)   on the appointment of an administrative receiver or a receiver or manager of its property under Chapter I of Part III of that Act, or the appointment of a receiver under Chapter II of that Part,
   (c)   on the passing of a resolution for voluntary winding-up without a declaration of solvency under section 89 of that Act, or
   (d)   on the making of a winding-up order under Part IV or V of that Act.

(3) For the purposes of this section a partnership becomes insolvent—
   (a)   on the making of a winding-up order against it under any provision of the Insolvency Act 1986 as applied by an order under section 420 of that Act, or
   (b)   when sequestration is awarded on the estate of the partnership under section 12 of the Bankruptcy (Scotland) Act 1985 or the partnership grants a trust deed for its creditors.

(4) For the purposes of this section an individual becomes insolvent—
   (a)   on the making of a bankruptcy order against him under Part IX of the Insolvency Act 1986, or
   (b)   on the sequestration of his estate under the Bankruptcy (Scotland) Act 1985 or when he grants a trust deed for his creditors.

(5) A company, partnership or individual shall also be treated as insolvent on the occurrence of any event corresponding to those specified in subsection (2), (3) or (4) under the law of Northern Ireland or of a country outside the United Kingdom.

(6) Where a provision is rendered ineffective by subsection (1), the parties are free to agree other terms for payment.

In the absence of such agreement, the relevant provisions of the Scheme for Construction Contracts apply.

Supplementary provisions

114   The Scheme for Construction Contracts

(1) The Minister shall by regulations make a scheme ("the Scheme for Construction Contracts") containing provision about the matters referred to in the preceding provisions of this Part.

(2) Before making any regulations under this section the Minister shall consult such persons as he thinks fit.

(3) In this section "the Minister" means—

    (a)   for England and Wales, the Secretary of State, and

    (b)   for Scotland, the Lord Advocate.

(4) Where any provisions of the Scheme for Construction Contracts apply by virtue of this Part in default of contractual provision agreed by the parties, they have effect as implied terms of the contract concerned.

(5) Regulations under this section shall not be made unless a draft of them has been approved by resolution of each House of Parliament.

115   Service of notices, &c

(1) The parties are free to agree on the manner of service of any notice or other document required or authorised to be served in pursuance of the construction contract or for any of the purposes of this Part.

(2) If or to the extent that there is no such agreement the following provisions apply.

(3) A notice or other document may be served on a person by any effective means.

(4) If a notice or other document is addressed, pre-paid and delivered by post—

    (a)   to the addressee's last known principal residence or, if he is or has been carrying on a trade, profession or business, his last known principal business address, or

    (b)   where the addressee is a body corporate, to the body's registered or principal office,

it shall be treated as effectively served.

(5) This section does not apply to the service of documents for the purposes of legal proceedings, for which provision is made by rules of court.

(6) References in this Part to a notice or other document include any form of communication in writing and references to service shall be construed accordingly.

116   Reckoning periods of time

(1) For the purposes of this Part periods of time shall be reckoned as follows.

(2) Where an act is required to be done within a specified period after or from a specified date, the period begins immediately after that date.

(3) Where the period would include Christmas Day, Good Friday or a day which under the Banking and Financial Dealings Act 1971 is a bank holiday in England and Wales or, as the case may be, in Scotland, that day shall be excluded.

117 Crown application

(1) This Part applies to a construction contract entered into by or on behalf of the Crown otherwise than by or on behalf of Her Majesty in her private capacity.

(2) This Part applies to a construction contract entered into on behalf of the Duchy of Cornwall notwithstanding any Crown interest.

(3) Where a construction contract is entered into by or on behalf of Her Majesty in right of the Duchy of Lancaster, Her Majesty shall be represented, for the purposes of any adjudication or other proceedings arising out of the contract by virtue of this Part, by the Chancellor of the Duchy or such person as he may appoint.

(4) Where a construction contract is entered into on behalf of the Duchy of Cornwall, the Duke of Cornwall or the possessor for the time being of the Duchy shall be represented, for the purposes of any adjudication or other proceedings arising out of the contract by virtue of this Part, by such person as he may appoint.

PART III

ARCHITECTS

The Architects Registration Board

118 The Board and its committees

(1) The Architects' Registration Council of the United Kingdom established under the Architects (Registration) Act 1931 ("the 1931 Act") shall be known as the Architects Registration Board.

(2) The Board of Architectural Education, the Admission Committee and the Discipline Committee constituted under the 1931 Act are abolished.

(3) In section 3 of the 1931 Act (constitution and functions of Architects' Registration Council), after subsection (2) insert—

"(2A) Part I of the First Schedule to this Act makes provision about the constitution and proceedings of the Board.

(2B) There shall be a Professional Conduct Committee of the Board and Part II of that Schedule makes provision about its constitution and proceedings.

(2C) Part III of that Schedule gives to the Board power to establish other committees and makes provision about their constitution and proceedings.

(2D) Part IV of that Schedule makes general provision about the Board and its committees.".

(4) For the First Schedule to the 1931 Act (constitution of Council) substitute the Schedule set out in Part I of Schedule 2.

119   Registrar and staff

For section 4 of the 1931 Act substitute—

"4 The Registrar

(1) The Board shall appoint a person to be known as the Registrar of Architects.

(2) The Board shall determine the period for which, and the terms on which, the Registrar is appointed.

(3) The Registrar shall have the functions provided by or by virtue of this Act and any other functions which the Board directs.

(4) The Board may, in addition to paying to the Registrar a salary or fees—

    (a)   pay pensions to or in respect of him or make contributions to the payment of such pensions; and

    (b)   pay him allowances, expenses and gratuities.

4A Staff

(1) The Board may appoint staff.

(2) The Board shall determine the period for which, and the terms on which, its staff are appointed.

(3) Staff appointed by the Board shall have the duties which the Board directs.

(4) The Board may, in addition to paying salaries to its staff—

    (a)   pay pensions to or in respect of them or make contributions to the payment of such pensions; and

    (b)   pay them allowances, expenses and gratuities.".

Registration and discipline

120   Registration

(1) Before section 6 of the 1931 Act insert—

"5A The Register

(1) The Registrar shall maintain the Register of Architects in which there shall be entered the name of every person entitled to be registered under this Act.

(2) The Register shall show the regular business address of each registered person.

(3) The Registrar shall make any necessary alterations to the Register and, in particular, shall remove from the Register the name of any registered person who has died or has applied in the prescribed manner requesting the removal of his name.

(4) The Board shall publish annually the current version of the Register and a copy of the most recently published version of the Register shall be provided to any person who requests one on payment of a reasonable charge determined by the Board.

(5) A copy of the Register purporting to be published by the Board shall be evidence (and, in Scotland, sufficient evidence) of any matter mentioned in it.

(6) A certificate purporting to be signed by the Registrar which states that a person—

    (a)   is registered;

    (b)   is not registered;

    (c)   was registered on a specified date or during a specified period;

    (d)   was not registered on a specified date or during a specified period; or

    (e)   has never been registered,

shall be evidence (and, in Scotland, sufficient evidence) of any matter stated.".

(2) For section 6 of the 1931 Act substitute—

"6 Entitlement to registration

(1) A person who has applied to the Registrar in the prescribed manner for registration in pursuance of this section is entitled to be registered if—

    (a)   he holds such qualifications and has gained such practical experience as may be prescribed; or

    (b)   he has a standard of competence which, in the opinion of the Board, is equivalent to that demonstrated by satisfying paragraph (a).

(2) The Board may require a person who applies for registration on the ground that he satisfies subsection (1)(b) to pass a prescribed examination in architecture.

(3) Before prescribing—

    (a)   qualifications or practical experience for the purposes of subsection (1)(a); or

    (b)   any examination for the purposes of subsection (2),

the Board shall consult the bodies representative of architects which are incorporated by royal charter and such other professional and educational bodies as it thinks appropriate.

(4) The Board may require—

    (a)   an applicant for registration in pursuance of this section; and

    (b)   a candidate for any examination under subsection (2),

to pay a fee of a prescribed amount.

(5) The Board may by rules prescribe the information and evidence to be furnished to the Registrar in connection with an application for registration in pursuance of this section.

(6) Where a person has duly applied for registration in pursuance of this section—

    (a)   if the Registrar is satisfied that the person is entitled to be registered, he shall enter his name in the Register; but

    (b)   if the Registrar is not so satisfied, he shall refer the application to the Board.

(7) The Registrar shall not consider an application for registration in pursuance of this section in any case in which it is inappropriate for him to do so (for instance because he is in any way connected with the applicant) but in such a case he shall refer the application to the Board.

(8) Where a person's application is referred to the Board under subsection (6) or (7), the Board shall direct the Registrar to enter the person's name in the Register if the Board is satisfied that the person is entitled to be registered.

(9) The Registrar shall serve on an applicant for registration in pursuance of this section written notice of the decision on his application—

    (a)   where the application is made on the ground that he satisfies subsection (1)(a), within three months of his application being duly made; and

    (b)   where the application is made on the ground that he satisfies subsection (1)(b), within six months of his application being duly made.".

(3) After section 6A of the 1931 Act insert—

"6B Retention of name in Register

(1) The Board may require a registered person to pay a fee (in this section referred to as a "retention fee") of a prescribed amount if he wishes his name to be retained in the Register in any calendar year after that in which it was entered.

(2) Where, after the Registrar has sent a registered person who is liable to pay a retention fee a written demand for the payment of the fee, the person fails to pay the fee within the prescribed period, the Registrar may remove the person's name from the Register.

(3) Where a person whose name has been removed from the Register under subsection (2) pays the retention fee, together with any further prescribed fee, before the end of the calendar year for which the retention fee is payable or such longer period as the Board may allow—

    (a)   his name shall be re-entered in the Register (without his having to make an application under section 6 or 6A); and

    (b)   if the Board so directs, it shall be treated as having been re-entered on the date on which it was removed.

6C Registration: additional requirements

(1) Where the Board is not satisfied that a person who—

    (a)   applies for registration in pursuance of section 6 or 6A;

    (b)   wishes his name to be retained or re-entered in the Register under section 6B; or

    (c)   applies for his name to be re-entered in the Register under section 7ZD,

has gained such recent practical experience as rules made by the Board require a person to have gained before he is entitled to have his name entered, retained or re-entered in the Register, his name shall not be so entered or re-entered, or shall be removed, unless he satisfies the Board of his competence to practise.

(2) Where the Board decides that the name of a person to whom paragraph (b) of subsection (1) applies is by virtue of that subsection to be removed from, or not to be re-entered in, the Register, the Registrar shall serve on him written notice of the decision within the prescribed period after the date of the decision.".

121 Discipline

For section 7 of the 1931 Act substitute—

"7 Unacceptable professional conduct and serious professional incompetence

(1) Where an allegation is made that a registered person is guilty of—

    (a) unacceptable professional conduct (that is, conduct which falls short of the standard required of a registered person); or

    (b) serious professional incompetence,

or it appears to the Registrar that a registered person may be so guilty, the case shall be investigated by persons appointed in accordance with rules made by the Board.

(2) Where persons investigating a case under subsection (1) find that a registered person has a case to answer they shall report their finding to the Professional Conduct Committee.

(3) Where the Professional Conduct Committee receives a report under subsection (2) in relation to a registered person, the Committee shall consider whether he is guilty of unacceptable professional conduct or serious professional incompetence.

(4) Before considering whether a registered person is guilty of unacceptable professional conduct or serious professional incompetence the Professional Conduct Committee shall—

    (a) serve on him written notice outlining the case against him; and

    (b) give him the opportunity to appear before the Committee to argue his case.

(5) At any such hearing the registered person is entitled to be legally represented.

(6) The Board may make rules as to the procedure to be followed by the Professional Conduct Committee in any proceedings under this section.

(7) If the Board does not make rules for the appointment of persons to investigate whether registered persons have been guilty of unacceptable professional conduct or serious professional incompetence, the Professional Conduct Committee shall consider such questions without any prior investigation.

7ZA Disciplinary orders

(1) The Professional Conduct Committee may make a disciplinary order in relation to a registered person if—

    (a)   it is satisfied, after considering his case, that he is guilty of unacceptable professional conduct or serious professional incompetence; or

    (b)   he has been convicted of a criminal offence other than an offence which has no material relevance to his fitness to practise as an architect.

(2) In this Act "disciplinary order" means—

    (a)   a reprimand;

    (b)   a penalty order;

    (c)   a suspension order; or

    (d)   an erasure order.

(3) Where the Professional Conduct Committee makes a disciplinary order in relation to a person, the Registrar shall serve written notice of the order on the person as soon as is reasonably practicable.

(4) The Professional Conduct Committee shall, at appropriate intervals and in such manner as it considers appropriate, publish—

    (a)   the names of persons whom it has found guilty of unacceptable professional conduct or serious professional incompetence or in relation to whom it has made a disciplinary order under subsection (1)(b); and

    (b)   in the case of each person a description of the conduct, incompetence or offence concerned and the nature of any disciplinary order made.

(5) Where, after considering the case of a registered person, the Professional Conduct Committee is not satisfied that he is guilty of unacceptable professional conduct or serious professional incompetence, it shall, if he so requests, publish a statement of that fact in such manner as it considers appropriate.

7ZB Penalty orders

(1) Where a penalty order is made in relation to a registered person, he shall pay to the Board the sum specified in the order.

(2) A penalty order may not specify a sum exceeding the amount which, at the relevant time, is the amount specified as level 4 on the standard scale of fines for summary offences.

In this subsection "the relevant time" means—

    (a)   in a case within subsection (1)(a) of section 7ZA, the time of the conduct or incompetence of which the registered person is found guilty; and

    (b)   in a case within subsection (1)(b) of that section, the time when he committed the criminal offence of which he has been convicted.

(3) A penalty order shall specify the period within which the sum specified in it is to be paid.

(4) If the person in relation to whom a penalty order is made does not pay the sum specified in the order within the period so specified, the Professional Conduct Committee may make a suspension order or an erasure order in relation to him.

(5) The Board shall pay into the Consolidated Fund any sum paid under a penalty order.

7ZC Suspension orders

Where a suspension order is made in relation to a registered person, the Registrar shall remove his name from the Register but shall re-enter it in the Register at the end of such period not exceeding two years as is specified in the order.

7ZD Erasure orders

(1) Where an erasure order is made in relation to a registered person, the Registrar shall remove his name from the Register and it shall not be re-entered in the Register unless the Board so directs.

(2) No application shall be made for the name of a person in relation to whom an erasure order has been made to be re-entered in the Register—

   (a) before the end of the period of two years beginning with the date of the erasure order or such longer period specified in the erasure order as the Professional Conduct Committee considers appropriate in a particular case; or

   (b) where he has made a previous application for his name to be re-entered in the Register, before the end of the prescribed period beginning with the date of the decision of the Board on that application.

(3) The Registrar shall serve on a person who applies for his name to be re-entered in the Register under this section written notice of the decision on his application within the prescribed period after the date of the decision.

(4) The Board may require a person whose name is re-entered in the Register under this section to pay a fee of a prescribed amount.".

122   Code of practice

After section 7ZD of the 1931 Act insert—

"7ZE Code of practice

(1) The Board shall issue a code laying down standards of professional conduct and practice expected of registered persons.

(2) The Board shall keep the code under review and vary its provisions whenever it considers it appropriate to do so.

(3) Before issuing or varying the code, the Board shall—

   (a) consult such professional bodies and such other persons with an interest in architecture as it considers appropriate; and

   (b) publish in such manner as it considers appropriate notice that it proposes to issue or vary the code, stating where copies of the proposals can be obtained.

(4) Failure by a registered person to comply with the provisions of the code—

(a)   shall not be taken of itself to constitute unacceptable professional conduct or serious professional incompetence on his part; but

(b)   shall be taken into account in any proceedings against him under section 7.

(5) The Board shall provide a copy of the code to any person who requests one on payment of a reasonable charge determined by the Board (and may provide a copy free of charge whenever it considers appropriate).".

Miscellaneous

123   Offence of practising while not registered

(1) In section 1 (prohibition on practising or carrying on business under title of architect by person who is not registered) of the Architects Registration Act 1938 ("the 1938 Act"), after subsection (1) insert—

"(1A) In this Act (and in section 17 of the principal Act) "business" includes any undertaking which is carried on for gain or reward or in the course of which services are provided otherwise than free of charge.".

(2) In section 3 of the 1938 Act (offence of practising while not registered), for the words from "to a fine" to "therefor:" substitute "to a fine not exceeding level 4 on the standard scale:".

(3) Re-number that section as subsection (1) of that section and after that subsection as so renumbered insert—

"(2) In relation to an offence under subsection (1)—

(a)   section 127(1) of the Magistrates' Courts Act 1980 (information to be laid within six months of offence);

(b)   Article 19(1) of the Magistrates' Courts (Northern Ireland) Order 1981 (complaint to be made within that time); and

(c)   section 136(1) of the Criminal Procedure (Scotland) Act 1995 (proceedings to be commenced within that time),

shall have effect as if for the references in them to six months there were substituted references to two years.".

(4) Re-number section 17 of the 1931 Act (defence for certain bodies corporate, firms and partnerships) as subsection (1) of that section and after that subsection as so renumbered insert—

"(2) The Board may by rules provide that subsection (1) shall not apply in relation to a body corporate, firm or partnership unless it has provided to the Board such information necessary for determining whether that subsection applies as may be prescribed.".

124   The Education Fund

(1) No fees received under the 1931 Act shall be credited to the Architects' Registration Council Education Fund ("the Fund") constituted under the Architects Registration (Amendment) Act 1969 ("the 1969 Act").

(2) The Board may transfer the assets of the Fund to such person and on such terms as may be approved by the Secretary of State.

(3) A person to whom the assets of the Fund are transferred under subsection (2) shall apply the assets, and all income arising from the assets, for the purposes authorised in subsection (4) of section 1 of the 1969 Act (assuming for this purpose that the reference in that subsection to the Council were a reference to the person to whom the assets of the Fund are transferred).

125 Supplementary

(1) The amendments made by Part II of Schedule 2, and the transitional provisions and savings in Part III of that Schedule, shall have effect.

(2) In this Part—

"the 1931 Act" means the Architects (Registration) Act 1931,

"the 1938 Act" means the Architects Registration Act 1938, and

"the 1969 Act" means the Architects Registration (Amendment) Act 1969.

(3) In this Part "the Fund" means the Architects' Registration Council Education Fund.

(4) The 1931 Act, the 1938 Act and this Part may be cited together as the Architects Acts 1931 to 1996.

PART IV

GRANTS &C. FOR REGENERATION, DEVELOPMENT AND RELOCATION

Financial assistance for regeneration and development

126 Power of Secretary of State to give financial assistance for regeneration and development

(1) The Secretary of State may, with the consent of the Treasury, give financial assistance to any person in respect of expenditure incurred in connection with activities which contribute to the regeneration or development of an area.

(2) Activities which contribute to the regeneration or development of an area include, in particular—

    (a) securing that land and buildings are brought into effective use;

    (b) contributing to, or encouraging, economic development;

    (c) creating an attractive and safe environment;

    (d) preventing crime or reducing the fear of crime;

    (e) providing or improving housing or social and recreational facilities, for the purpose of encouraging people to live or work in the area or of benefiting people who live there;

    (f) providing employment for local people;

    (g) providing or improving training, educational facilities or health services for local people;

(h)   assisting local people to make use of opportunities for education, training or employment;

(i)   benefiting local people who have special needs because of disability or because of their sex or the racial group to which they belong.

(3) In subsection (2)—

"local people", in relation to an area, means people who live or work in the area; and

"racial group" has the same meaning as in the Race Relations Act 1976.

127   Regeneration and development: forms of assistance

(1) Financial assistance under section 126 (powers of Secretary of State to give financial assistance) may be given in any form.

(2) Assistance may, in particular, be given by way of—

(a)   grants,

(b)   loans,

(c)   guarantees, or

(d)   incurring expenditure for the benefit of the person assisted.

(3) The Secretary of State must not, in giving financial assistance under section 126, purchase loan or share capital in a company.

128   Regeneration and development: terms on which assistance is given

(1) Financial assistance under section 126 may be given on such terms as the Secretary of State, with the consent of the Treasury, considers appropriate.

(2) The terms may, in particular, include provision as to—

(a)   circumstances in which the assistance is to be repaid, or otherwise made good, to the Secretary of State, and the manner in which that is to be done; or

(b)   circumstances in which the Secretary of State is entitled to recover the proceeds or part of the proceeds of any disposal of land or buildings in respect of which assistance was provided.

(3) The person receiving assistance must comply with the terms on which it is given, and compliance may be enforced by the Secretary of State.

129   Regeneration and development: consequential amendment

In section 175(2)(b) of the Leasehold Reform, Housing and Urban Development Act 1993, for the words from "sections 27 to 29" to the end, substitute "sections 126 to 128 of the Housing Grants, Construction and Regeneration Act 1996 (financial assistance for regeneration and development)".

130   Regeneration and development: Welsh Development Agency

(1) In the Welsh Development Agency Act 1975, after section 10 insert—

"10A Financial assistance for regeneration and development

(1) The Secretary of State may appoint the Agency to act as his agent in connection with such of his functions mentioned in subsection (2) below as he may specify.

(2) The functions are—

(a) functions under sections 126 to 128 of the Housing Grants, Construction and Regeneration Act 1996 (financial assistance for regeneration and development), so far as they relate to—

(i) financial assistance which the Agency has power to give apart from this section; or

(ii) financial assistance given under that Act in pursuance of an agreement entered into by the Secretary of State for Wales before the coming into force of this section, or

(b) functions of the Secretary of State in relation to financial assistance given by the Secretary of State for Wales under sections 27 to 29 of the Housing and Planning Act 1986.

(3) An appointment under this section shall be on such terms as the Secretary of State, with the approval of the Treasury, may specify; and the Agency shall act under the appointment in accordance with those terms.

(4) The Agency's powers in relation to functions under an appointment under this section include the powers it has in relation to functions under subsection (3) of section 1 by virtue of subsections (6) and (7) of that section."

(2) In section 2(8) of that Act, after "declared that" insert ", except as provided by section 10A below,".

Relocation grants in clearance areas

131   Resolution by local housing authority to pay relocation grants

(1) Before deciding whether to declare an area to be a clearance area under section 289 of the Housing Act 1985, a local housing authority shall—

(a) consider whether their resources are sufficient for the purpose of carrying into effect a resolution declaring the power to pay relocation grants to be exercisable as regards that area; and

(b) in deciding that question, have regard to such guidance as may from time to time be given by the Secretary of State.

(2) Where a local housing authority decide that their resources are sufficient for that purpose, they shall—

(a) consider whether to pass such a resolution; and

(b) notify every person on whom notice is required to be served under subsection (2B)(a) of section 289 of the Housing Act 1985 that they are so considering and invite him to make representations.

(3) In deciding whether to pass such a resolution, a local housing authority shall—

   (a) have regard to such guidance as may from time to time be given by the Secretary of State; and

   (b) take account of any representations made by persons notified under subsection (2)(b).

(4) Where a local housing authority pass such a resolution, they shall transmit a copy of it to the Secretary of State at the same time as they transmit to him a copy of the resolution under section 289 of the Housing Act 1985.

(5) Subsections (2) to (4) of section 604A of the Housing Act 1985 (duty to consider guidance before taking enforcement action) shall apply in relation to guidance under subsection (1)(b) or (3)(a) as they apply in relation to guidance under subsection (1) of that section.

132 Relocation grants: applications and payments

(1) Where a local housing authority have passed a resolution declaring the power to pay relocation grants to be exercisable as regards a clearance area, they may pay such grants for the purpose of enabling qualifying persons to acquire qualifying dwellings (see section 133).

(2) No relocation grant shall be paid unless—

   (a) an application for it is made to the authority by a qualifying person in accordance with the provisions of this section and is approved by them;

   (b) the application is accompanied by a certificate falling within subsection (5) in respect of the qualifying dwelling to which the application relates; and

   (c) such other conditions (whether as to the dwelling or the interest to be acquired or otherwise) as may be prescribed are fulfilled,

and regulations made under paragraph (c) may provide for particular questions arising under the regulations to be determined by the authority.

(3) An application for a relocation grant shall be in writing and shall specify the qualifying dwelling to which it relates and contain such particulars as may be prescribed.

(4) The Secretary of State may by regulations prescribe a form of application for a relocation grant and an application to which any such regulations apply shall not be validly made unless it is in the prescribed form.

(5) A certificate under this subsection certifies—

   (a) that the applicant proposes to acquire an owner's interest in the qualifying dwelling to which the application relates; and

   (b) that he, or a member of his family, intends to live in that dwelling as his (or that member's) only or main residence throughout the grant condition period.

(6) A relocation grant shall be paid in such manner and at such time as the authority may determine having regard to the purpose for which it is paid.

(7) Nothing in section 25 of the Local Government Act 1988 (consent required for provision of financial assistance) shall apply in relation to any exercise of the power to pay relocation grants.

133 Relocation grants: qualifying persons and qualifying dwellings

(1) A person is a qualifying person for the purposes of section 132 (relocation grants: applications and payments) if—

    (a) an interest of his in a dwelling in the clearance area ("the original dwelling") has been, or is to be, acquired by the local housing authority under section 290 of the Housing Act 1985 or section 154 of the Town and Country Planning Act 1990;

    (b) that interest on the acquisition date was greater than a tenancy for a year or from year to year; and

    (c) the original dwelling was his only or main residence both on the declaration date and throughout the period of 12 months ending with the acquisition date.

(2) A dwelling is a qualifying dwelling for the purposes of section 132 if it is—

    (a) in the clearance area; or

    (b) in an area designated by the local housing authority as an area for the relocation of persons displaced by the clearance;

and any area so designated may be in or outside the authority's area.

(3) In making a designation under subsection (2) a local housing authority shall have regard to such guidance as may from time to time be given by the Secretary of State.

(4) Subsections (2) to (4) of section 604A of the Housing Act 1985 (duty to consider guidance before taking enforcement action) shall apply in relation to guidance under subsection (3) as they apply in relation to guidance under subsection (1) of that section.

(5) Any reference in the preceding provisions of this section to the clearance area includes a reference to any land surrounded by or adjoining the clearance area which has been, or is to be, acquired by the local housing authority under section 290 of the Housing Act 1985 or section 154 of the Town and Country Planning Act 1990.

(6) In this section—

> "the acquisition date", in relation to an acquisition under section 290 of the Housing Act 1985, means the date of—
>
>   (a) the notice to treat under section 5 of the Compulsory Purchase Act 1965;
>
>   (b) the general vesting declaration under section 4 of the Compulsory Purchase (Vesting Declarations) Act 1981; or
>
>   (c) the agreement between the local housing authority and the applicant,
>
> in pursuance of which the interest in the original dwelling was, or is to be, acquired by the authority;
>
> "the acquisition date", in relation to an acquisition under section 154 of the Town and Country Planning Act 1990 (effect of valid blight notice), means the date mentioned in subsection (3) of that section;
>
> "the declaration date" means the date on which the clearance area was declared by the authority.

134 Relocation grants: amount

(1) Subject to subsections (2) to (4), the amount of any relocation grant shall be such amount as the local housing authority may determine.

(2) The amount of any relocation grant shall not exceed such amount as may be prescribed.

(3) The amount of any relocation grant shall not exceed the difference between—

    (a)   the cost of acquiring the qualifying dwelling to which the application relates; and

    (b)   such part as may be prescribed of the amount which has been, or is to be, paid by the authority in respect of the acquisition of the applicant's interest in the original dwelling.

(4) If the financial resources of the applicant exceed the applicable amount, the amount of any grant which may be paid shall, in accordance with regulations, be reduced from what it would otherwise have been.

(5) For the purposes of subsection (3), the cost of acquiring the qualifying dwelling shall be taken to be whichever of the following is the lesser amount, namely—

    (a)   the actual cost (including reasonable incidental expenses) of acquiring the dwelling; and

    (b)   the amount which the authority considers to be the reasonable cost (including such expenses) of acquiring a comparable dwelling in the same area.

(6) Provision may be made by regulations—

    (a)   for the determination of the amount which is to be taken to be the financial resources of an applicant,

    (b)   for the determination of the applicable amount referred to in subsection (4), and

    (c)   as to circumstances in which the financial resources of an applicant are to be assumed (by reason of his receiving a prescribed benefit or otherwise) not to exceed the applicable amount.

(7) Regulations may, in particular—

    (a)   make provision for account to be taken of the income, assets, needs and outgoings not only of the applicant himself but also of his spouse, any person living with him or intending to live with him and any person on whom he is dependent or who is dependent on him;

    (b)   make provision for amounts specified in or determined under the regulations to be taken into account for particular purposes.

(8) Regulations may apply, subject to such modifications as may be prescribed by the regulations, any other statutory means-testing regime as it has effect from time to time.

(9) Regulations may make provision requiring any information or evidence needed for the determination of any matter under this section to be furnished by such person as may be prescribed.

(10) In this section—

    "the original dwelling" has the same meaning as in section 133;

    "regulations" means regulations made by the Secretary of State with the consent of the Treasury.

135   Relocation grants: condition for repayment on disposal

(1) It is a condition of a relocation grant that, if an owner of the qualifying dwelling makes a relevant disposal (other than an exempt disposal) of the dwelling within the grant

condition period, he shall repay to the local housing authority on demand the amount of the grant.

(2) A condition under this section is binding on any person who is for the time being an owner of the qualifying dwelling.

(3) Where the authority have the right to demand repayment of an amount as mentioned in subsection (1), they may—

    (a)   if the case falls within subsection (4), or

    (b)   in any other case, with the consent of the Secretary of State,

determine not to demand payment or to demand a lesser amount.

(4) The cases referred to in subsection (3)(a) are where the authority are satisfied that the owner of the dwelling—

    (a)   is elderly or infirm and is making the disposal with the intention—

        (i) of going to live in a hospital, hospice, sheltered housing, residential care home or similar institution as his only or main residence, or

        (ii) of moving to somewhere where care will be provided by any person; or

    (b)   is making the disposal with the intention of going to live with and care for an elderly or infirm member of his family or his partner's family.

(5) The consent of the Secretary of State for the purposes of subsection (3)(b) may be given either generally or in relation to any one or more specified authorities or descriptions of authority or in relation to particular cases or descriptions of case.

(6) A condition under this section shall cease to be in force with respect to a dwelling if there is a relevant disposal of the dwelling that is an exempt disposal, other than—

    (a)   a disposal within section 54(1)(a) (disposal to associates of person making disposal), or

    (b)   a disposal within section 54(1)(b) (vesting under will or on intestacy) to a person who resided with the deceased in the dwelling as his only or main residence throughout the period of twelve months ending with the date of the deceased's death.

(7) Any disposal which—

    (a)   by virtue of section 53 (meaning of relevant disposal), is a relevant disposal; or

    (b)   by virtue of section 54 (meaning of exempt disposal), is an exempt disposal,

for the purposes of the provisions of Part I of this Act (relating to grant conditions) is also such a disposal for the purposes of this section.

136   Relocation grants: conditions as to owner-occupation

(1) It is a condition of a relocation grant that throughout the grant condition period the qualifying dwelling is occupied in accordance with the intention stated in the certificate under section 132(5)(b).

(2) It is also a condition of the grant that if at any time when that condition is in force the local housing authority serve notice on the owner of the qualifying dwelling requiring him to do so, he will within the period of 21 days beginning with the date on which the notice was served furnish to the authority a statement showing how that condition is being fulfilled.

(3) A condition under this section is binding on any person who is for the time being an owner of the dwelling.

(4) In the event of a breach of a condition under this section, the owner for the time being of the dwelling shall on demand repay to the local housing authority the amount of the grant, together with compound interest on that amount as from the beginning of the grant condition period, calculated at such reasonable rate as the authority may determine and with yearly rests.

(5) The local housing authority may determine not to make such a demand or to demand a lesser amount.

(6) Subsections (6) and (7) of section 135 apply for the purposes of this section as they apply for the purposes of that section.

137 Relocation grants: cessation of conditions on repayment of grant, &c

(1) If at any time while a condition under section 135 or 136 (a "grant condition") remains in force with respect to a qualifying dwelling—

    (a) the owner of the dwelling to which the condition relates pays the amount of the grant to the local housing authority by whom the grant was made, or

    (b) a mortgagee of the interest of the owner in that dwelling being a mortgagee entitled to exercise a power of sale, makes such a payment, or

    (c) the local housing authority determine not to demand repayment on the breach of a grant condition, or

    (d) the authority demand repayment in whole or in part on the breach of a grant condition and that demand is satisfied,

the grant condition and any other grant conditions shall cease to be in force with respect to that dwelling.

(2) An amount paid by a mortgagee under subsection (1)(b) shall be treated as part of the sums secured by the mortgage and may be discharged accordingly.

138 Relocation grants: liability to repay is a charge on dwelling

(1) The liability that may arise under a condition under section 135, or under section 136(4), is a charge on the qualifying dwelling, taking effect as if it had been created by deed expressed to be by way of legal mortgage.

(2) The charge has priority immediately after any legal charge securing an amount—

    (a) advanced to the applicant by an approved lending institution for the purpose of enabling him to acquire the dwelling, or

    (b) further advanced to him by that institution;

but the local housing authority may at any time by written notice served on an approved lending institution postpone the charge taking effect by virtue of this section to a legal charge securing an amount advanced or further advanced to the applicant by that institution.

(3) A charge taking effect by virtue of this section is a land charge for the purposes of section 59 of the Land Registration Act 1925 notwithstanding subsection (5) of that section (exclusion of mortgages), and subsection (2) of that section applies accordingly with respect to its protection and realisation.

(4) A condition under section 135 or 136 does not, by virtue of its binding any person who is for the time being an owner of the dwelling, bind a person exercising rights under a charge having priority over the charge taking effect by virtue of this section, or a person deriving title under him.

(5) The approved lending institutions for the purposes of section 156 of the Housing Act 1985 (right to buy: liability to repay discount is a charge on premises) are also approved lending institutions for the purposes of this section.

139   Relocation grants: contributions by the Secretary of State

(1) The Secretary of State may pay contributions to local housing authorities towards such expenditure incurred by them under section 132 (payment of relocation grants) as he may determine.

(2) The rate or rates of the contributions, the calculation of the expenditure to which they relate and the manner of their payment shall be such as may be determined by the Secretary of State.

(3) Any determination under subsection (1) or (2) may be made generally, or with respect to a particular local housing authority or description of authority, including a description framed by reference to authorities in a particular area.

(4) Contributions under this section shall be payable subject to such conditions as to repayment, and such conditions as to records, certificates, audit or otherwise, as the Secretary of State may impose.

140   Minor definitions relating to relocation grants

(1) In sections 131 to 139 (provisions as to relocation grants)—

"dwelling" means a building or part of a building occupied or intended to be occupied as a separate dwelling, together with any yard, garden, outhouses and appurtenances belonging to it or usually enjoyed with it;

"grant condition period" means the period of five years, or such other period as the Secretary of State may by order specify, beginning with the date of the acquisition of the owner's interest in the qualifying dwelling;

"local housing authority" has the same meaning as in the Housing Act 1985;

"owner", in relation to a dwelling, means the person who—

(a) is for the time being entitled to receive from a lessee of the dwelling (or would be so entitled if the dwelling were let) a rent of not less than two-thirds of the net annual value of the dwelling; and

(b) is not himself liable as lessee of the dwelling, or of property which includes the dwelling, to pay such a rent to a superior landlord;

"owner's interest", in relation to any premises, means—

(a) an estate in fee simple absolute in possession, or

(b) a term of years absolute of which not less than five years remain unexpired at the date of the application,

whether held by the applicant alone or jointly with others;

"prescribed" means prescribed by regulations made by the Secretary of State;

"qualifying dwelling" has the meaning given by section 133(2);

"qualifying person" has the meaning given by section 133(1);

"relocation grant" means a grant under section 132.

(2) For the purposes of the definition of "owner" in subsection (1), the net annual value of a dwelling means the rent at which the dwelling might reasonably be expected to be let from year to year if the tenant undertook to pay all usual tenant's rates and taxes and to bear the cost of repair and insurance and the other expenses, if any, necessary to maintain the dwelling in a state to command that rent.

(3) Any dispute arising as to the net annual value of a dwelling shall be referred in writing for decision by the district valuer.

In this subsection "district valuer" has the same meaning as in the Housing Act 1985.

(4) Section 113 of the Housing Act 1985 (meaning of "members of a person's family") applies in determining whether a person is a member of another's family for the purposes of sections 132 and 135.

PART V

MISCELLANEOUS AND GENERAL PROVISIONS

Miscellaneous provisions

141   Existing housing grants: meaning of exempt disposal

(1) Section 124 of the Local Government and Housing Act 1989 (relevant and exempt disposals for purposes of housing grants) is amended as follows.

(2) In subsection (3) (exempt disposals), for paragraph (c) substitute—

"(c)   a disposal of the whole of the dwelling in pursuance of any such order as is mentioned in subsection (4A) below;".

(3) After subsection (4) insert—

"(4A) The orders referred to in subsection (3)(c) above are orders under—

(a)   section 24 or 24A of the Matrimonial Causes Act 1973 (property adjustment orders or orders for the sale of property in connection with matrimonial proceedings),

(b)   section 2 of the Inheritance (Provision for Family and Dependants) Act 1975 (orders as to financial provision to be made from estate),

(c)   section 17 of the Matrimonial and Family Proceedings Act 1984 (property adjustment orders or orders for the sale of property after overseas divorce, &c.), or

(d)   paragraph 1 of Schedule 1 to the Children Act 1989 (orders for financial relief against parents);".

142 Home energy efficiency schemes

(1) In section 15 of the Social Security Act 1990 (grants for the improvement of energy efficiency in certain dwellings, &c.) for subsection (1) (power to make grants) substitute—

"(1) The Secretary of State may make or arrange for the making of grants—

    (a) towards the cost of carrying out work for the purpose of—

        (i) improving the thermal insulation of dwellings, or

        (ii) otherwise reducing or preventing the wastage of energy in dwellings (whether in connection with space or water heating, lighting, the use of domestic appliances or otherwise), and

    (b) where any such work is, or is to be, carried out, towards the cost of providing persons with advice on reducing or preventing the wastage of energy in dwellings;

but no grants shall be made under this section except in accordance with regulations made by the Secretary of State.".

(2) In subsection (10) of that section, after the definition of "functions", insert—

""materials" includes space and water heating systems;".

143 Urban development corporations: pre-dissolution transfers

(1) After section 165A of the Local Government, Planning and Land Act 1980 insert—

"165B Transfer of property, rights and liabilities to statutory bodies

(1) Subject to this section, the Secretary of State may at any time by order transfer to a statutory body, upon such terms as he thinks fit, any property, rights or liabilities which—

    (a) are for the time being vested in an urban development corporation, and

    (b) are not proposed to be transferred under section 165 or 165A above.

(2) An order under this section may terminate—

    (a) any appointment of the corporation under subsection (1) of section 177 of the Leasehold Reform, Housing and Urban Development Act 1993 (power of corporations to act as agents of the Urban Regeneration Agency); and

    (b) any arrangements made by the corporation under subsection (2) of that section.

(3) An order under this section may—

    (a) establish new bodies corporate to receive any property, rights or liabilities to be transferred by an order under this section;

    (b) amend, repeal or otherwise modify any enactment for the purpose of enabling any body established under any enactment to receive such property, rights or liabilities.

(4) An order under this section—

(a) may contain such incidental, consequential, transitional or supplementary provision as the Secretary of State thinks necessary or expedient (including provisions amending, repealing or otherwise modifying any enactment); and

(b) shall be made by statutory instrument which shall be subject to annulment in pursuance of a resolution of either House of Parliament.

(5) Before making an order under this section, the Secretary of State shall consult each local authority in whose area all or part of the urban development area is situated.

(6) In this section—

"enactment" includes any instrument made under any enactment;

"statutory body" means any body established under this section or any other enactment.".

(2) In consequence of the above amendment, the Local Government, Planning and Land Act 1980 is amended as follows.

(3) In section 165(9) (meaning of local authority) for "sections 165A and 166" substitute "sections 165A to 166".

(4) In section 165A(1) (power of Secretary of State to transfer property &c. to himself) for paragraph (b) substitute—

"(b) are not proposed to be transferred under section 165 above or 165B below.".

(5) In section 166(5) (dissolution of corporations) after "section 165A" insert "or 165B".

144 Housing action trusts: orders for dissolution

(1) Section 88 of the Housing Act 1988 (dissolution of housing action trusts) is amended as follows.

(2) In subsection (4) (contents of dissolution orders) after paragraph (a) insert—

"(aa) where it provides for any such disposal or transfer as is mentioned in subsection (2)(b) above, may contain provisions—

(i) establishing new bodies corporate to receive the disposal or transfer; or

(ii) amending, repealing or otherwise modifying any enactment for the purpose of enabling any body established under any enactment to receive the disposal or transfer;".

(3) In paragraph (b) of that subsection (supplementary and transitional provisions) for the words from "any enactment", where it first appears, to "order" substitute ", repealing or otherwise modifying any enactment".

(4) After that subsection insert—

"(5) In this section "enactment" includes any instrument made under any enactment.".

145 The Commission for the New Towns: orders for dissolution

(1) In Schedule 9 to the New Towns Act 1981 (additional provisions as to the Commission for the New Towns) paragraph 7 (power to dissolve Commission) is amended as follows.

(2) After sub-paragraph (2) insert—

> "(2A) Any order under this paragraph may—
>
> > (a) establish new bodies corporate to receive any property, rights, liabilities or obligations vested by an order under this paragraph;
> >
> > (b) amend, repeal or otherwise modify any enactment for the purpose of enabling any body established under any enactment to receive such property, rights, liabilities or obligations.".

(3) In sub-paragraph (3) (incidental, supplemental, consequential or transitional provision) for the words from "amendments" to the end substitute "provisions amending, repealing or otherwise modifying any enactment.".

(4) For sub-paragraph (7) (interpretation) substitute—

> "(7) In this paragraph—
>
> > "accountable public authority" means a body established under this paragraph or any other enactment;
> >
> > "enactment" includes any instrument made under any enactment.".

General provisions

146 Orders, regulations and directions

(1) Orders, regulations and directions under this Act may make different provision for different cases or descriptions of case, including different provision for different areas.

(2) Orders and regulations under this Act may contain such incidental, supplementary or transitional provisions and savings as the Secretary of State considers appropriate.

(3) Orders and regulations under this Act shall be made by statutory instrument which, except for—

> (a) orders and regulations subject to affirmative resolution procedure (see sections 104(4), 105(4), 106(4) and 114(5)),
>
> (b) orders under section 150(3), or
>
> (c) regulations which only prescribe forms or particulars to be contained in forms,

shall be subject to annulment in pursuance of a resolution of either House of Parliament.

147 Repeals and revocations

The enactments specified in Schedule 3 are repealed or revoked to the extent specified.

148 Extent

(1) The provisions of this Act extend to England and Wales.

(2) The following provisions of this Act extend to Scotland—

Part II (construction contracts),

Part III (architects),

sections 126 to 128 (financial assistance for regeneration and development), and

Part V (miscellaneous and general provisions), except—

    (i) sections 141, 144 and 145 (which amend provisions which do not extend to Scotland), and

    (ii) Part I of Schedule 3 (repeals consequential on provisions not extending to Scotland).

(3) The following provisions of this Act extend to Northern Ireland—

Part III (architects), and

Part V (miscellaneous and general provisions), except—

    (i) sections 142 to 145 (home energy efficiency schemes and residuary bodies), and

    (ii) Parts I and III of Schedule 3 (repeals consequential on provisions not extending to Northern Ireland).

(4) Except as otherwise provided, any amendment or repeal by this Act of an enactment has the same extent as the enactment amended or repealed.

149   Corresponding provision for Northern Ireland

An Order in Council under paragraph 1(1)(b) of Schedule 1 to the Northern Ireland Act 1974 (legislation for Northern Ireland in the interim period) which states that it is made only for purposes corresponding to those of Part II (construction contracts) or section 142 (home energy efficiency schemes)—

    (a) shall not be subject to paragraph 1(4) and (5) of that Schedule (affirmative resolution of both Houses of Parliament), but

    (b) shall be subject to annulment in pursuance of a resolution of either House of Parliament.

150   Commencement

(1) The following provisions of this Act come into force on Royal Assent—

section 146 (orders, regulations and directions),

sections 148 to 151 (extent, commencement and other general provisions).

(2) The following provisions of this Act come into force at the end of the period of two months beginning with the date on which this Act is passed—

sections 126 to 130 (financial assistance for regeneration and development),

section 141 (existing housing grants: meaning of exempt disposal),

section 142 (home energy efficiency schemes),

sections 143 to 145 (residuary bodies),

Part III of Schedule 3 (repeals consequential on Part IV) and section 147 so far as relating to that Part.

(3) The other provisions of this Act come into force on a day appointed by order of the Secretary of State, and different days may be appointed for different areas and different purposes.

(4) The Secretary of State may by order under subsection (3) make such transitional provision and savings as appear to him to be appropriate in connection with the coming into force of any provision of this Act.

151 Short title

This Act may be cited as the Housing Grants, Construction and Regeneration Act 1996.

SCHEDULES

SCHEDULE 1 Section 103.

PRIVATE SECTOR RENEWAL: CONSEQUENTIAL AMENDMENTS

Rent Act 1977 (c. 42)

1   (1) Section 116 of the Rent Act 1977 (court order where tenant unwilling to consent to works) is amended as follows.

(2) In subsection (2), omit "any of paragraphs (a) to (c) of".

(3) For subsection (3) substitute—

> "(3) The condition is that the works were specified in an application for a renovation grant, a common parts grant, a disabled facilities grant or an HMO grant under Chapter I of Part I of the Housing Grants, Construction and Regeneration Act 1996 and the application has been approved.".

(4) In subsection (5), for the words from "under section 512(2)" to the end, substitute "under section 37 of the Housing Grants, Construction and Regeneration Act 1996.".

Housing Act 1985 (c. 68)

2   In section 47(4) of the Housing Act 1985 (limitation of service charges: deduct amount of grant), for the words from "Part XV" to "or conversion)" substitute "section 523 of the Housing Act 1985 (assistance for provision of separate service pipe for water supply) or any provision of Part I of the Housing Grants, Construction and Regeneration Act 1996 (grants, &c. for renewal of private sector housing) or any corresponding earlier enactment".

3   In section 48(3A) of the Housing Act 1985 (information as to relevant costs: grant), for the words from "Part XV" to the end substitute "section 523 of the Housing Act 1985 (assistance for provision of separate service pipe for water supply) or any provision of Chapter I of Part I of the Housing Grants, Construction and Regeneration Act 1996 (grants for renewal of private sector housing) or any corresponding earlier enactment".

4   (1) In section 100(2) of the Housing Act 1985 (power to reimburse cost of tenant's improvements; grant), for "improvement grant" to "Part XV" substitute "renovation grant or common parts grant under Chapter I of Part I of the Housing Grants, Construction and Regeneration Act 1996 (grants for renewal of private sector housing)".

(2) In that section, omit subsection (2A).

5   (1) In section 101(1) of the Housing Act 1985 (rent not to be increased on account of tenant's improvements: grant), for "improvement grant" to the end substitute "renovation grant or common parts grant under Chapter I of Part I of the Housing

Grants, Construction and Regeneration Act 1996 (grants for renewal of private sector housing).".

(2) In that section, omit subsection (1A).

6   In section 190A of the Housing Act 1985 (repair notices and group repair schemes) —

     (a)   in subsection (2), for "subsection 130(1)" to the end substitute "subsection 66(1) of the Housing Grants, Construction and Regeneration Act 1996).".

     (b)   in subsection (5), for "Part VIII" to the end substitute "Chapter II of Part I of the Housing Grants, Construction and Regeneration Act 1996 (group repair schemes).".

7   (1) In section 244 of the Housing Act 1985 (environmental works: no assistance where grant made), for subsection (3) substitute—

     "(3) No such assistance shall be given towards works in respect of which an application for renovation grant or common parts grant under Chapter I of Part I of the Housing Grants, Construction and Regeneration Act 1996 (grants for renewal of private sector housing) has been approved.".

(2) In that section, omit subsection (3A).

8   (1) In subsection (2)(b) of section 255 of the Housing Act 1985 (general powers of local housing authority not to include making grants), for "an improvement grant" to the end substitute "a renovation grant or common parts grant might be made under Chapter I of Part I of the Housing Grants, Construction and Regeneration Act 1996 (grants for renewal of private sector housing).".

(2) In that section, omit subsection (3).

9   In section 535(1)(a) of the Housing Act 1985 (exclusion of assistance under Part XV of that Act where grant application pending or approved), for the words from "an improvement grant" to "Part XV" substitute "renovation grant or common parts grant under Chapter I of Part I of the Housing Grants, Construction and Regeneration Act 1996 (grants for renewal of private sector housing).".

10   In section 605 of the Housing Act 1985 (consideration by local housing authority of housing conditions in their district), for subsection (1)(e) substitute—

     "(e)   Part I of the Housing Grants, Construction and Regeneration Act 1996 (grants, &c. for renewal of private sector housing).".

Landlord and Tenant Act 1985 (c. 70)

11   (1) In subsection (1) of section 20A of the Landlord and Tenant Act 1985 (limitation of service charges: grant-aided works), for the words from "Part XV" to "conversion)" substitute "section 523 of the Housing Act 1985 (assistance for provision of separate service pipe for water supply) or any provision of Part I of the Housing Grants, Construction and Regeneration Act 1996 (grants, &c. for renewal of private sector housing) or any corresponding earlier enactment".

(2) In subsection (2) of that section—

     (a)   for "Part VIII of the Local Government and Housing Act 1989" substitute "Part I of the Housing Grants, Construction and Regeneration Act 1996"; and

      (b)   for "the outstanding balance determined in accordance with subsections (3) and (4) of section 130 of that Act" substitute "the balance of the cost determined in accordance with section 69(3) of the Housing Grants, Construction and Regeneration Act 1996".

12   In section 21 of the Landlord and Tenant Act 1985 (request for summary of relevant costs)—

      (a)   in subsection (5), for the words from "Part XV" to "conversion)" substitute "section 523 of the Housing Act 1985 (assistance for provision of separate service pipe for water supply) or any provision of Part I of the Housing Grants, Construction and Regeneration Act 1996 (grants, &c. for renewal of private sector housing) or any corresponding earlier enactment"; and

      (b)   in subsection (5B) for "Part VIII of the Local Government and Housing Act 1989" substitute "Chapter II of Part I of the Housing Grants, Construction and Regeneration Act 1996 or any corresponding earlier enactment".

Housing Act 1988 (c. 50)

13   In section 121(1) of the Housing Act 1988 (rent officers' functions), for "section 110" to the end substitute "section 31 of the Housing Grants, Construction and Regeneration Act 1996 applies.".

Local Government and Housing Act 1989 (c. 42)

14   In section 93(5) of the Local Government and Housing Act 1989 (general powers of local housing authority: works in renewal area), for "Part VIII of this Act" substitute "Part I of the Housing Grants, Construction and Regeneration Act 1996".

15   (1) Section 169 of the Local Government and Housing Act 1989 (power of local authority and Secretary of State to provide professional, &c. services in relation to works) is amended as follows.

    (2) In subsection (2)(b), for "section 114(3) or (4) above" substitute "section 23 of the Housing Grants, Construction and Regeneration Act 1996 (disabled facilities grants: purposes)".

    (3) In subsection (2)(c), for "section 115(3) above" substitute "or under section 12 or 27 of the Housing Grants, Construction and Regeneration Act 1996 (renovation grants or HMO grants: purposes)".

    (4) For subsection (2)(d) substitute—

        "(d)   works in relation to home repair assistance under sections 76 to 79 of the Housing Grants, Construction and Regeneration Act 1996.".

SCHEDULE 2

ARCHITECTS

PART I

NEW FIRST SCHEDULE TO THE 1931 ACT

1 This is the Schedule to be substituted for the First Schedule to the 1931 Act—

"FIRST
SCHEDULE

THE BOARD AND ITS COMMITTEES

PART I

THE BOARD

Membership

Sections 118 and 125.

Section 3.

1 The Board shall consist of—
 (a) seven elected members; and
 (b) eight appointed members.

Elected members

2 (1) The elected members shall be elected in accordance with an electoral scheme made by the Board, with the approval of the Privy Council, after consultation with such bodies as appear to the Board to be representative of architects.

 (2) An electoral scheme under sub-paragraph (1) may be amended by the Board with the approval of the Privy Council and after consultation with such bodies as are mentioned in that sub-paragraph.

 (3) The persons qualified—
  (a) to elect the elected members; and
  (b) to be elected as elected members,
 are all those who are registered persons when the election is held.

Appointed members

3 (1) The appointed members shall be appointed by the Privy Council, after consultation with the Secretary of State and such other persons or bodies as the Privy Council think fit, to represent the interests of users of architectural services and the general public.

 (2) No registered person shall be eligible for appointment as an appointed member.

Term of office

4   (1) Subject to sub-paragraphs (2) and (3), the term of office of a member of the Board is three years.

(2) A member may resign at any time by notice in writing addressed to the Registrar.

(3) The Board may by rules prescribe grounds (such as repeated absence from meetings or unacceptable professional conduct) on which any member may be removed from office and the procedure for removal.

5   A person who has held office as a member of the Board for a continuous period of six years may not be elected or appointed as a member until at least three years have elapsed since he last held office.

Casual vacancies

6   (1) Where a vacancy occurs among the members of the Board otherwise than by the expiry of a member's term of office—
    (a)   if the vacancy is among the elected members, the Board shall appoint a registered person to fill it; and
    (b)   if the vacancy is among the appointed members, the Privy Council shall appoint a person to fill it.

(2) Subject to paragraph 4(2) and (3), a person appointed under sub-paragraph (1) to fill a vacancy holds office until the date on which the term of office of the member whose vacancy he fills would have expired.

(3) A person appointed under sub-paragraph (1)(a) shall be regarded as an elected member and a person appointed under sub-paragraph (1)(b) shall be regarded as an appointed member.

Chairman

7   (1) The members of the Board shall elect a chairman from among themselves.

(2) The chairman—
    (a)   may resign by notice in writing addressed to the Registrar; and
    (b)   may be removed by a majority vote of the other members of the Board.

(3) Rules made by the Board may make provision for the appointment of a person to act as chairman in the event of a vacancy in the office of chairman or in such other circumstances as may be prescribed.

8   In the event of a tie in any vote of the Board the chairman shall have an additional casting vote.

Procedure

9   The quorum of the Board shall be nine, of whom at least four shall be elected members and at least four shall be appointed members.

10   The Board may make rules governing its meetings and procedure.

PART II

THE PROFESSIONAL CONDUCT COMMITTEE

11   The Professional Conduct Committee shall consist of—

    (a)   four elected members of the Board, including at least one whose address in the Register is in Scotland, or (if there is no elected member whose address in the Register is in Scotland or no such elected member who is willing to act) three elected members and one registered person whose address in the Register is in Scotland;

    (b)   three appointed members of the Board; and

    (c)   two persons nominated by the President of the Law Society.

12   (1) The members of the Professional Conduct Committee shall elect a chairman from among themselves.

    (2) The chairman—

    (a)   may resign by notice in writing addressed to the Registrar; and

    (b)   may be removed by a majority vote of the other members of the Professional Conduct Committee.

    (3) Rules made by the Board may make provision for the appointment of a person to act as chairman in the event of a vacancy in the office of chairman or in such other circumstances as may be prescribed.

13   (1) The quorum of the Professional Conduct Committee shall be one elected member of the Board, one appointed member of the Board and one person nominated by the President of the Law Society.

    (2) Where the Committee is considering the case of a person whose address in the Register is in Scotland, the Committee is not quorate unless there is present a member of the Committee who is a registered person and whose address in the Register is in Scotland.

14   In the event of a tie in any vote of the Professional Conduct Committee the chairman shall have an additional casting vote; and in any proceedings relating to a registered person the additional vote shall be cast in favour of that person.

15   The Board may make rules governing the selection and term of office of members of the Professional Conduct Committee (including casual vacancies).

PART III

OTHER COMMITTEES

16   The Board may establish such committees as it considers appropriate to discharge any of its functions under this Act other than—

    (a)   prescribing fees under section 6(4), 6A(1A), 6B(1) or (3) or 7ZD(4); or

    (b)   acting under section 6(1), (2) or (5), 6A(1) or (1B), 6C(1) or 7ZE(1), (2) or (3),

or to assist the Board in the discharge by the Board of any of its functions.

17   (1) Any committee established by the Board may include persons who are not members of the Board; but if a committee is established to discharge any function of the Board, the majority of the members of the committee must be members of the Board.

    (2) Subject to that, the membership of any committee established by the Board shall be determined by the Board.

18   No vote of any committee established by the Board for the discharge of any of its functions shall be valid unless the majority of those voting are members of the Board.

19   The Board may make rules governing the term of office of members of any committee established by the Board (including casual vacancies) and the meetings and procedure (including chairmanship and quorum) of any committee established by the Board.

Part IV

General

20   (1) The Board, the Professional Conduct Committee and any committee established by the Board may exercise its functions even though there is a vacancy among its members.

    (2) No proceedings of the Board, the Professional Conduct Committee or any committee established by the Board are invalidated by any defect in the election or appointment of a member.

21   The Board may by rules provide for the payment to members of the Board, the Professional Conduct Committee or any committee established by the Board of—

    (a)   fees for attendance at meetings of the Board or committee; and

    (b)   travelling and subsistence allowances in respect of attendance at such meetings or the conduct of business of the Board or committee.

22   (1) The Secretary of State may, after consultation with the Board and such other persons or bodies as he thinks fit, by order amend the provisions of this Schedule.

    (2) An order under sub-paragraph (1) shall be made by statutory instrument which shall be subject to annulment in pursuance of a resolution of either House of Parliament.".

PART II

OTHER AMENDMENTS

The 1931 Act

2   (1) Section 2 of the 1931 Act (interpretation) is amended as follows.

(2) For the definition of "the Council" substitute—

"The expression "the Board" means the Architects Registration Board.".

(3) In the definition of "registered person", for "registered under this Act" substitute "whose name is in the Register".

(4) In the definition of "prescribed", for "regulations made by the Council" substitute "rules made by the Board".

(5) In the definition of "the Register", for "kept in pursuance of this Act" substitute "of Architects".

(6) After that definition insert—

"The expression "the Registrar" means the Registrar of Architects appointed by the Board under section 4.

The expressions "penalty order", "suspension order" and "erasure order" shall be construed in accordance with sections 7ZB, 7ZC and 7ZD.

The expression "disciplinary order" has the meaning given by section 7ZA.".

3   (1) Section 3 of the 1931 Act (constitution and functions of Architects' Registration Council) is amended as follows.

(2) In subsection (1), for the words from the beginning to "name," substitute "The Architects Registration Board shall be a body corporate".

(3) In subsection (2)—
    (a)   omit the first sentence, and
    (b)   in the second sentence, for "Council" (in both places) substitute "Board".

(4) For the sidenote substitute "The Board and its committees.".

4   (1) Section 6A of the 1931 Act (European qualifications) is amended as follows.

(2) In subsection (1), for the words from "shall" to the end substitute "and has applied to the Registrar in the prescribed manner for registration in pursuance of this section is entitled to be registered.".

(3) After that subsection insert—

"(1A) The Board may require an applicant for registration in pursuance of this section to pay a fee of a prescribed amount.

(1B) The Board may by rules prescribe the information and evidence to be furnished to the Registrar in connection with an application for registration in pursuance of this section.".

(4) In subsection (7), for the words from "The Council" to "aware that" substitute "An application by a person for registration in pursuance of this section may be refused if".

(5) In subsection (8), for the words from "Council" to "applicant" substitute "Registrar shall serve on an applicant for registration in pursuance of this section written notice of the decision on his application".

(6) In subsection (9)—

    (a)   for "Council consult" substitute "Board consults", and

    (b)   for "Council of" substitute "Board of".

5   (1) Section 7A of the 1931 Act (removal of name from Register: disqualification in another member State) is amended as follows.

(2) In subsection (1)—

    (a)   for "Council were" substitute "Board was",

    (b)   for "Council, on" substitute "Board, on", and

    (c)   for "cause his name to be removed" substitute "order the Registrar to remove his name".

(3) In subsection (2), for "7 of this Act" substitute "7ZA(1)".

(4) After that subsection insert—

    "(3) Where the Board orders the Registrar to remove a person's name from the Register under this section, the Registrar shall serve written notice of the removal on the person as soon as is reasonably practicable.".

6   (1) Section 9 of the 1931 Act (right of appeal against removal from Register) is amended as follows.

(2) For the words "by the removal" onwards substitute "by—

    (a)   his name not being re-entered in, or being removed from, the Register by virtue of section 6C(1);

    (b)   the making of a disciplinary order in relation to him; or

    (c)   the Board ordering the Registrar to remove his name from the Register under section 7A,

may appeal to the High Court or the Court of Session within three months from the date on which notice of the decision or order concerned is served on him; and on an appeal under this section the Court may make any order which appears appropriate, and no appeal shall lie from any decision of the Court on such an appeal.".

(3) For the sidenote substitute "Appeals.".

7   In section 11 of the 1931 Act (removal of name from Register for failure to notify change of address), for "Council" (in each place) substitute "Registrar".

8   In section 12 of the 1931 Act (penalty for obtaining registration by false representation), for "wilfully" substitute "intentionally".

9   For section 13 of the 1931 Act (regulations) substitute—

"13 Rules

"13   "13 Rules

(1) The Board may make rules generally for carrying out or facilitating the purposes of this Act.

(2) The Board shall, before making any rules under this Act, publish a draft of the rules and give those to whom the rules would be applicable an opportunity of making representations to the Board.".

10   (1) Section 15 of the 1931 Act (supply of regulations and forms) is amended as follows.

(2) For "Council" substitute "Registrar".

(3) For "regulations" (in each place, including the sidenote) substitute "rules".

11   (1) Section 16 of the 1931 Act (service of documents) is amended as follows.

(2) In subsection (1), for "to be sent" substitute "to be served".

(3) In subsection (2), for "to the removal from the Register of the name of any registered person" substitute "required to be served by section 6C(2), 7(4)(a), 7ZA(3) or 7A(3)".

12   (1) Section 17 of the 1931 Act (defence for certain bodies corporate, firms and partnerships) is amended as follows.

(2) In paragraph (a), for "superintendent who is a registered person and" substitute "registered person".

(3) In paragraph (b), for the words from "and" to "who is" substitute "it is carried on by or under the supervision of".

(4) For the sidenote substitute "Defence for business under control and management of registered person.".

13   For section 18(2) of the 1931 Act (application to Northern Ireland) substitute—

"(2) This Act extends to Northern Ireland.".

The 1938 Act

14   (1) Section 1A of the 1938 Act (visiting EC architects) is amended as follows.

(2) In subsections (2), (3) and (4), for "Council" (in each place) substitute "Registrar".

(3) In subsection (3), for "they consider" substitute "the Registrar considers".

(4) In subsection (6), for the words from "when" to the end substitute "when—

    (a)   he is subject to a disqualifying decision in another member State;

    (b)   his name has been removed from the Register pursuant to a suspension order or an erasure order and has not been re-entered; or

    (c)   he is required under section 6C(1) of the principal Act to satisfy the Board of his competence to practise but has not done so.".

(5) In subsection (8), for the words from the beginning to "the regulation of" substitute "The provisions of, and of rules under, the principal Act relating to".

15   In section 3 of the 1938 Act (offence of practising while not registered), in the proviso—

(a)   in paragraph (a), omit "of the Council" and "subsection (2) of section six of", and

(b)   for paragraphs (b) and (c) substitute—

"(b)   in a case where the contravention is occasioned by the removal of the defendant's name from the Register in circumstances in which notice is required to be served on him—

(i) that the notice had not been duly served before that date,

(ii) that the time for bringing an appeal against the removal had not expired at that date, or

(iii) that such an appeal had been duly brought, but had not been determined, before that date.".

16   In section 5 of the 1938 Act (construction and citation), in subsection (2), for the words from "Acts 1931 and" to the end substitute "Act 1931".

17   For section 6(1) of the 1938 Act (application to Northern Ireland) substitute—

"(1) This Act extends to Northern Ireland.".

Other enactments

18   In section 6 of the Inspection of Churches Measure 1955 (interpretation), in the definition of "qualified person", for "Architects Registration Acts 1931 to 1969" substitute "Architects Acts 1931 to 1996".

19   In section 52(1) of the Cathedrals Measure 1963 (interpretation), in the definition of "architect", for "Architects (Registration) Acts 1931 to 1938" substitute "Architects Acts 1931 to 1996".

20   In section 20(1) of the Care of Cathedrals Measure 1990 (interpretation), in the definition of "architect", for "Architects Registration Acts 1931 to 1969" substitute "Architects Acts 1931 to 1996".

PART III

TRANSITIONAL PROVISIONS AND SAVINGS

First elections and appointments to the Board

21   (1) Part I of the First Schedule to the 1931 Act as substituted by Part I of this Schedule shall have effect before the appointed day so far as is necessary to enable the election and appointment of members of the Board to take office on that day.

(2) Until the appointed day references to the Board in paragraph 2 of that Schedule shall have effect as references to the Council.

(3) Where persons elected or appointed as members of the Board by virtue of this paragraph attend meetings before the appointed day in preparation for the conduct of business of the Board on or after that day, the Council may pay to them any such

fees or travelling or subsistence allowances in respect of their attendance as appear appropriate.

(4) The term of office of the members of the Board appointed by the Privy Council (by virtue of this paragraph) to take office on the appointed day—

(a) is one year beginning with that day in the case of three of those members,

(b) is two years beginning with that day in the case of another three of those members, and

(c) is three years beginning with that day in the case of the remaining two members.

Registration

22 Where before the appointed day a person has duly applied for registration under the 1931 Act but no decision on the application has been made, the application shall be dealt with on and after the appointed day in the same way as an application duly made on or after that day (except that no further fee may be required to be paid).

23 Examinations in architecture which immediately before the appointed day were recognised by the Council for the purposes of subsection (1)(c) of section 6 of the 1931 Act (as it has effect before the substitution made by section 120 of this Act) shall (subject to rules made by the Board) be treated on and after that day as qualifications prescribed under subsection (1)(a) of that section (as it has effect after that substitution).

24 Section 6B of the 1931 Act shall have effect as if the reference in subsection (3) of that section to a person whose name has been removed from the Register under subsection (2) of that section included a reference to a person whose name was removed from the Register under section 13(5) of the 1931 Act before the appointed day.

25 The first reference to the Board in section 7A(1) of the 1931 Act shall be construed, in relation to the entry of a name in the Register at a time before the appointed day, as a reference to the Council.

Discipline

26 Where before the appointed day—

(a) the Discipline Committee has begun an inquiry into a case in which it is alleged that a registered person has been guilty of conduct disgraceful to him in his capacity as an architect, but

(b) the Council has not decided whether to remove his name from the Register,

the case shall be referred to the Professional Conduct Committee which shall consider whether he is guilty of unacceptable professional conduct or serious professional incompetence.

27 (1) Subject to sub-paragraph (2), the provisions substituted by section 121 of this Act for section 7 of the 1931 Act have effect in relation to anything done or omitted to be done before the appointed day as in relation to anything done or omitted to be done after that day.

(2) The Professional Conduct Committee—

(a) may only make a disciplinary order in respect of anything done, or omitted to be done, by a person before the appointed day if the Council could have removed his name from the Register under section 7 of the 1931 Act (as it had effect before the substitution made by section 121 of this Act), and

(b)   may not make a reprimand or penalty order in respect of anything so done or omitted to be so done.

Pre-commencement removals and disqualifications

28   (1) Where a person's name has been removed from the Register under section 7 of the 1931 Act before the appointed day, he may at any time on or after that day apply to the Board for his name to be re-entered in the Register.

(2) If he does so, the Board may direct that his name shall be re-entered in the Register.

(3) The Registrar shall serve on a person who applies for his name to be re-entered in the Register under this paragraph written notice of the decision on his application within the prescribed period after the date of the decision.

(4) The Board may require a person whose name is re-entered in the Register under this paragraph to pay a fee of such amount, not exceeding the fee then payable by an applicant for registration in pursuance of section 6 of the 1931 Act, as may be prescribed.

29   A person may appeal under section 9 of the 1931 Act against—

(a)   the removal of his name from the Register before the appointed day, or

(b)   a determination of the Council before the appointed day that he be disqualified for registration during any period,

within three months from the date on which notice of the removal or determination was served on him.

30   Section 1A(6)(b) of the 1938 Act shall have effect as if it included a reference to a period of disqualification imposed by the Council.

Offence of practising while not registered

31   The amendments made in sections 1 and 3 of the 1938 Act and section 17 of the 1931 Act by section 123(1), (3) and (4) of this Act do not apply in relation to an offence committed before the appointed day.

32   The repeal made in section 3 of the 1938 Act by section 123(2) of this Act applies in relation to an offence committed before the appointed day (as well as in relation to one committed on or after that day).

Transfer of Fund

33   If the transfer of the assets of the Fund takes place after the appointed day, the repeal by this Act of sections 1(1) and (4) to (6), 3 and 4 of the 1969 Act shall not come into force until the transfer is made; and during the period beginning with the appointed day and ending with the transfer references in those provisions to the Council shall have effect as references to the Board.

Supplementary

34   (1) In this Part of this Schedule—

(a)   "the Board" means the Architects Registration Board, and

(b)   other expressions used in the 1931 Act have the same meanings as in that Act.

(2) In this Part of this Schedule "appointed day" means the day appointed by the Secretary of State for the coming into force of this Part of this Act.

35   Nothing in this Schedule prejudices the operation of section 16 or 17 (effect of repeals) of the Interpretation Act 1978.

SCHEDULE 3

Section 147.

REPEALS AND REVOCATIONS

PART I

GRANTS, &C FOR RENEWAL OF PRIVATE SECTOR HOUSING

Chapter	*Short title*	*Extent of repeal*
1977 c. 42.	Rent Act 1977.	In section 116(2), the words "any of paragraphs (a) to (c) of".
1985 c. 68.	Housing Act 1985.	Section 100(2A).
		Section 101(1A).
		Section 244(3A).
		Section 255(3).
1989 c. 42.	Local Government and Housing Act 1989.	Part VIII.
		In Schedule 11, paragraph 52, paragraph 63, and paragraphs 66 to 69.
1993 c. 10.	Charities Act 1993.	In Schedule 6, paragraph 30, the words "The Local Government and Housing Act 1989 section 138(1)".
1994 c. 19.	Local Government (Wales) Act 1994.	In Schedule 8, paragraph 10(1) and (2).
1994 c. 29.	Police and Magistrates' Courts Act 1994.	In Schedule 4, paragraph 40.

PART II

ARCHITECTS

Chapter	Short title	Extent of repeal
21 & 22 Geo.5 c. 33.	Architects (Registration) Act 1931.	In section 3, in subsection (2), the first sentence and subsections (3) and (4).
		Section 5.
		In section 6A(1), the words "Subject to the provisions of this Act,".
		In section 7A(1), the words "of this Act".
		Section 8.
		The Second Schedule.
		The Third Schedule.
1 & 2 Geo.6 c. 54.	Architects Registration Act 1938.	In section 1(3), the words "the words "Registered Architects" in subsection (3) of section three of the principal Act, and for", "respectively" and "the word "Architects" and".
		In section 3, the words "of the Council" and "subsection (2) of section six of".
1969 c. 42.	Architects Registration (Amendment) Act 1969.	The whole Act.
1977 c. 45.	Criminal Law Act 1977.	In Schedule 6, the entry relating to the Architects Registration Act 1938.
S.I. 1987/1824.	Architects' Qualifications (EEC Recognition) Order 1987.	Article 4.
1995 c. 40.	Criminal Procedure (Consequential Provisions) (Scotland) Act 1995.	In Schedule 2, in Part II, the entry relating to the Architects Registration Act 1938.

PART III

FINANCIAL ASSISTANCE FOR REGENERATION AND DEVELOPMENT

Chapter	*Short title*	*Extent of repeal*
1986 c. 63.	Housing and Planning Act 1986.	Part III. In section 58(1) and (2), the words "Part III (financial assistance for urban regeneration);".
1993 c. 28.	Leasehold Reform, Housing and Urban Development Act 1993.	Section 174. In section 188(6), the words "174,".